CLIMATE CHANGE AND ITS CAUSES, EFFECTS AND PREDICTION

ROLE OF EXPERIMENTAL FORESTS AND RANGES IN UNDERSTANDING AND ADAPTING TO CLIMATE CHANGE

CLIMATE CHANGE AND ITS CAUSES, EFFECTS AND PREDICTION

Additional books in this series can be found on Nova's website under the Series tab.

Additional e-books in this series can be found on Nova's website under the e-book tab.

CLIMATE CHANGE AND ITS CAUSES, EFFECTS AND PREDICTION

ROLE OF EXPERIMENTAL FORESTS AND RANGES IN UNDERSTANDING AND ADAPTING TO CLIMATE CHANGE

KAYLA PIERCE
EDITOR

New York

Copyright © 2015 by Nova Science Publishers, Inc.

All rights reserved. No part of this book may be reproduced, stored in a retrieval system or transmitted in any form or by any means: electronic, electrostatic, magnetic, tape, mechanical photocopying, recording or otherwise without the written permission of the Publisher.

For permission to use material from this book please contact us:
nova.main@novapublishers.com

NOTICE TO THE READER

The Publisher has taken reasonable care in the preparation of this book, but makes no expressed or implied warranty of any kind and assumes no responsibility for any errors or omissions. No liability is assumed for incidental or consequential damages in connection with or arising out of information contained in this book. The Publisher shall not be liable for any special, consequential, or exemplary damages resulting, in whole or in part, from the readers' use of, or reliance upon, this material. Any parts of this book based on government reports are so indicated and copyright is claimed for those parts to the extent applicable to compilations of such works.

Independent verification should be sought for any data, advice or recommendations contained in this book. In addition, no responsibility is assumed by the publisher for any injury and/or damage to persons or property arising from any methods, products, instructions, ideas or otherwise contained in this publication.

This publication is designed to provide accurate and authoritative information with regard to the subject matter covered herein. It is sold with the clear understanding that the Publisher is not engaged in rendering legal or any other professional services. If legal or any other expert assistance is required, the services of a competent person should be sought. FROM A DECLARATION OF PARTICIPANTS JOINTLY ADOPTED BY A COMMITTEE OF THE AMERICAN BAR ASSOCIATION AND A COMMITTEE OF PUBLISHERS.

Additional color graphics may be available in the e-book version of this book.

Library of Congress Cataloging-in-Publication Data

ISBN: 978-1-63463-729-9

Published by Nova Science Publishers, Inc. † *New York*

CONTENTS

Preface   **vii**

Chapter 1   Experimental Forests and Climate Change: Views
of Long-Term Employees on Ecological Change
and the Role of Experimental Forests and Ranges
in Understanding and Adapting to Climate Change   **1**
Laurie Yung, Mason Bradbury
and Daniel R. Williams

Chapter 2   Climate Change Science: Key Points   **75**
Jane A. Leggett

Index   **95**

PREFACE

For more than a century, Experimental Forests and Ranges (EFRs) have provided critical science on the ecosystems and management activities of the National Forest System. Forest Service EFRs play a unique and important role both within the agency as well as in the broader field of land management. The goal of EFRs is to generate knowledge that benefits both public land managers and private land owners. This goal is achieved through research projects on pressing natural resource topics such as hydrology, fire dynamics, range management, erosion, climate change, silviculture, and forest regeneration. EFRs are uniquely situated for such research due to their relative stability and long-term datasets. This book specifically discusses the role of EFRs in understanding and adapting to climate change.

Chapter 1 – In this project, the authors examined the views of 21 long-term employees on climate change in 14 Rocky Mountain Research Station Experimental Forests and Ranges (EFRs). EFRs were described by employees as uniquely positioned to advance knowledge of climate change impacts and adaptation strategies due to the research integrity they provide for long-term studies, the ability to host experimental treatments on the efficacy of adaptation actions, and the opportunity for long-term field observations to inform and improve research. Institutional commitment and capacity was identified by participants as critical to realizing the potential of EFRs to contribute to climate change research.

Chapter 2 – Though climate change science often is portrayed as controversial, broad scientific agreement exists on many points: The Earth's climate is warming and changing. Human-related emissions of greenhouse gases (GHG) and other pollutants have contributed to warming observed since the 1970s and, if continued, would tend to drive further warming, sea level rise, and other climate shifts. Volcanoes, the Earth's relationship to the Sun, solar cycles, and land cover change may be more influential on climate shifts than rising GHG concentrations on other time and geographic scales. Human-induced changes are super-imposed on and interact with natural climate variability. The largest uncertainties in climate projections surround feedbacks in the Earth system that augment or dampen the initial influence, or affect the pattern of changes. Feedback mechanisms are apparent in clouds, vegetation, oceans, and potential emissions from soils. There is a wide range of projections of future, human-induced climate change, all pointing toward warming and associated sea level rise, with wider uncertainties regarding the nature of precipitation, storms, and other important aspects of climate. Human societies and ecosystems are sensitive to climate. Some past climate changes benefited civilizations; others contributed to the demise of some societies. Small future changes may bring benefits for some and adverse effects to others. Large climate changes would be increasingly adverse for a widening scope of populations and ecosystems. As is common and constructive in science, scientists debate finer points. For example, a large majority but not all scientists find compelling evidence that rising GHG have contributed the most influence on global warming since the 1970s, with solar radiation a smaller influence on that time scale. Most climate modelers project important impacts of unabated GHG emissions, with low likelihoods of catastrophic impacts over this century. Human influences on climate change would continue for centuries after atmospheric concentrations of GHG are stabilized, as the accumulated gases continue to exert effects and as the Earth's systems seek to equilibrate. The U.S. government and others have invested billions of dollars in research to improve understanding of the Earth's climate system, resulting in major improvements in understanding while major uncertainties remain. However, it is fundamental to the scientific method that science does not provide absolute proofs; all scientific theories are to some degree provisional and may be rejected or modified based on new evidence. Private and public decisions to act or not to act, to reduce the human contribution to climate change or to prepare for future changes, will be made in the context of accumulating evidence (or lack of evidence), accumulating GHG concentrations, ongoing debate about risks, and other considerations (e.g., economics and distributional effects).

In: Role of Experimental Forests and Ranges ...   ISBN: 978-1-63463-729-9
Editor: Kayla Pierce   © 2015 Nova Science Publishers, Inc.

Chapter 1

EXPERIMENTAL FORESTS AND CLIMATE CHANGE: VIEWS OF LONG-TERM EMPLOYEES ON ECOLOGICAL CHANGE AND THE ROLE OF EXPERIMENTAL FORESTS AND RANGES IN UNDERSTANDING AND ADAPTING TO CLIMATE CHANGE[*]

Laurie Yung, Mason Bradbury and Daniel R. Williams

ABSTRACT

In this project, we examined the views of 21 long-term employees on climate change in 14 Rocky Mountain Research Station Experimental Forests and Ranges (EFRs). EFRs were described by employees as uniquely positioned to advance knowledge of climate change impacts and adaptation strategies due to the research integrity they provide for long-term studies, the ability to host experimental treatments on the efficacy of adaptation actions, and the opportunity for long-term field observations to inform and improve research. Institutional commitment and capacity was identified by participants as critical to realizing the potential of EFRs to contribute to climate change research.

---

[*] This is an edited, reformatted and augmented version of a research paper, RMRS-RP-100, issued by the United States Department of Agriculture, Forest Service, Rocky Mountain Research Station, October 2012.

EXECUTIVE SUMMARY

For more than a century, Experimental Forests and Ranges (EFRs) have provided critical science on the ecosystems and management activities of the National Forest System. Through in-depth interviews with 21 long-term EFR employees and retirees, we examined the potential role of the 14 Rocky Mountain Research Station (RMRS) EFRs in building knowledge of climate change impacts and advancing climate change adaptation. Interviews also explored ecological change on EFRs and the role of informal field observations in the scientific process.

Research participants described EFRs as dynamic ecosystems and outlined a variety of specific changes, including shifts in species composition, non-native species invasion, changes to disturbance processes, and insect and disease outbreaks.

Anthropogenic climate change was believed to be causing the decline of certain species, conversion to new habitat types, increases in pathogens, and shifts in the timing of wildlife activities. Many participants acknowledged the influence of both natural cycles and variability and the effects of anthropogenic climate change. These changes are not unique to EFRs; they are shifts that are being observed throughout the Interior West.

According to research participants, what is unique is the ability of EFRs to contribute to our understanding of ecological change and the efficacy of management actions. EFRs represent most ecosystems in the Interior West, and they build on a history of relevance and science leadership. With regard to EFRs and climate change, the following opportunities and challenges were identified:

- EFRs provide the research integrity necessarily for long-term studies; climate change research requires a long-term approach.
- Long-term data collection, one of the hallmarks of EFRs, is particularly important to understanding climate change impacts and adaptation actions. Effective long-term research is enabled by research integrity, the potentially unparalleled level of control and protection that EFR designation provides.
- EFRs enable manipulative research; research on the efficacy of management interventions for climate change adaptation requires experimental treatments.
- Because EFRs allow manipulative, experimental research at multiple scales, they are uniquely positioned to examine the impacts of specific

management actions over the short and long term, which is critical given the largely untested nature of proposed climate change adaptation actions.

- Long-term relationships between scientists and EFRs improve research quality; field observations over longer time-scales can improve climate change research.
- Informal field observations help generate relevant hypotheses and aid in interpretation of results.
- Realizing the potential of EFRs to contribute to climate change research requires institutional commitment and capacity, and long-term investment.

Research participants described the ways in which institutional change (largely related to organizational structure, allocation of funding, and research priorities) influenced the ability of EFRs to contribute to understanding the impacts of climate change.

Participants argued that the opportunities outlined above can only be realized through long-term commitment and investment, including adequate funding to maintain sites and long-term data collection, and making difficult decisions regarding how to allocate funds in a budget shortfall. In the context of limited funds, RMRS can also build capacity at EFRs through strategic partnerships and participation in relevant data networks.

PROJECT GOALS

- Document the long-term and short-term ecological changes observed by employees at USDA Forest Service Experimental Forests and Ranges in the Interior West and gain an understanding of how these changes might be related to climate change.
- Examine the role that USDA Forest Service Experimental Forests and Ranges might play in understanding climate change impacts and advancing climate change adaptation in the Interior West.

RESEARCH QUESTIONS

- What changes have long-term employees at EFRs observed over the last 10 to 40 years? What are the perceived causes of those changes? What changes are attributed specifically to climate change?
- What roles do formal data collection and experiential, informal observation play in understanding such changes and developing experiments for the forest?
- How can an understanding of change on EFRs contribute to our knowledge of climate change impacts? What role can EFRs play in advancing our knowledge of the efficacy of different approaches to climate change adaptation?

USDA FOREST SERVICE EXPERIMENTAL FORESTS AND RANGES

Forest Service EFRs play a unique and important role both within the agency as well as in the broader field of land management. In total, the Forest Service operates 80 EFRs, 74 of which are in the continental United States, with an additional 4 in Alaska, 2 in Hawaii, and 1 in Puerto Rico (Wells and others 2009). EFRs vary in size from Kawishiwi Experimental Forest in Minnesota, at 47 ha, to Desert Experimental Range in Utah, at 22,500 ha (Adams and others 2008). The idea for institutionalized EFRs originated at an 1868 convention of German foresters and soil scientists in Vienna (Young 2008). Bernard Fernow, a German forester who immigrated to the United States, introduced the idea to U.S. foresters when he became the head of the U.S. Division of Forestry in 1886 (Young 2008). This idea became reality when Coconino Experiment Station (Coconino later became Fort Valley Experimental Forest) was established in 1908 under the direction of Gifford Pinchot and Raphael Zon, a former student of Fernow's (Young 2008). The number of EFRs grew to 7 by 1915 (Young 2008) and 29 by the 1930s (Wells and others 2009).

Since their inception, EFRs have been sites for long-term research and demonstration of findings in forest and range management (Adams and others 2008). The goal of EFRs is to generate knowledge that benefits both public land managers and private land owners (Wells and others 2009). This goal is achieved through research projects on pressing natural resource topics such as hydrology, fire dynamics, range management, erosion, climate change, silviculture, and forest regeneration (Wells and others 2009). EFRs are uniquely situated for such research due to their relative stability and long-term datasets (Adams and others 2008). In recent years, EFRs have become increasingly connected to each other and to external collaborators through regional and national research networks (Mowrer 2010). Through such collaborations, as well as their long-term datasets and national distribution, EFRs are effective venues for monitoring and understanding ecological change from local to global scales.

Figure 1. Map of EFRs (green triangles, brown square, and blue inverted triangle) in the Rocky Mountain Region.

There are 14 EFRs under the Rocky Mountain Research Station: 3 in Arizona, 2 in Utah, 2 in Colorado, 1 in Wyoming, 3 in Idaho, 2 in Montana, and 1 in South Dakota (Adams and others 2008). These include several notable EFRs, such as Fort Valley Experimental Forest in Arizona, the first Experimental Forest established, and Desert Experimental Range, Coram

Experimental Forest, and Fraser Experimental Forest, which were declared Biosphere Reserves in 1976 by the United Nations Education Scientific and Cultural Organization's Man and Biosphere Program (Adams and others 2008). EFRs in the Interior West include a spectrum of vegetation types, from desert to alpine (Mowrer 2010), and host research on a wide array of topics, such as ponderosa pine ecology, hydrology, and fire dynamics (Adams and others 2008). More detailed descriptions of RMRS EFRs can be found in Appendices A and B.

Climate Change Impacts

EFRs are already helping us better understand the regional and ecosystem-specific impacts of global climate change. Following, we briefly outline some of the current or anticipated climate change impacts in the Interior West. Global climate change and the altered temperatures and precipitation regimes that characterize it are likely to affect species and ecosystems in several important ways in the Interior West (Samberg 2011a). Effects include: changes in species distributions, phenology (Parmesan and Yohe 2003), impacts from invasive species (Samberg 2011c), and disturbance regimes (Samberg 2011d). There is strong evidence that species and ecosystems are already experiencing some of these effects (Parmesan and Yohe 2003).

One ecological impact of climate change that ecosystems are experiencing is the shifting of historical species ranges. The general trend of this shift is pole-ward and upward in elevation, with species showing expansion of high latitude and high-elevation ranges and contraction of low latitude and low elevation ranges (Parmesan 2006). Parmesan, in her extensive review of ecological responses to climate change, describes such a trend in penguins, which have moved south toward the South Pole over the past 20 to 50 years, and United Kingdom songbirds, which have moved 18.9 km north over the past 20 years. Additionally, trees in the Canadian Rockies and Siberia and lowland birds in Costa Rica have shifted their ranges upward in elevation (Parmesan 2006). Meanwhile, many lower elevation pika (*Ochotona princeps*) populations in the Great Basin have disappeared (Parmesan 2006).

In addition to altering species ranges, climate change affects phenology (Parmesan 2006). Phenological shifts are occurring in plants in the form of changed bud burst and first flowering dates, and in animals, in changed nesting and mating dates. In Japan, records dating back to 1400 show that, after varying with no apparent trends for 500 years, the date when cherry blossoms first appear each season has become significantly earlier in the last 100 years (Parmesan 2006). Egg laying dates for 65 species of United Kingdom birds moved 8.8 days earlier from 1971 to 1995 (Parmesan 2006). Importantly, the responses of phenology to climate change are not identical across species (Parmesan 2006), meaning that climate change has the potential to disassociate biotic communities (Stephenson and others 2010).

Another ecological impact of climate change is its potential to increase the threat and number of invasive species (Samberg 2011c). Climate change may facilitate the spread of invasive species through the removal of temperature and precipitation constraints on species and the creation of open ecological niches (Samberg 2011c). Furthermore, because invasive species are adapted to rapid change, through short generation times, high dispersal capabilities, and wide environmental tolerances, they may be better at coping with a rapidly changing climate (Samberg 2011c). Climate change may also necessitate a reassessment or refining of our definitions of invasive (Stephenson and others 2010), as it will become difficult to distinguish between non-invasive non-natives and invasive non-natives because of shifting species distributions (Samberg 2011c).

Climate change is also likely to alter ecosystem disturbances, including fire and flood regimes (Stephenson and others 2010) and windstorms (Frelich and Reich 2009). In the case of fire regimes, climate change is projected to increase fire intensity, frequency, and area burned (Samberg 2011d). Such an alteration in fire regimes will favor fire-tolerant species, replace mixed-age stands with even-aged stands, and, in some locations, catalyze shifts to new ecosystem states (Samberg 2011d).

An issue especially important to the Interior West is the potential for the combined effects of environmental stressors, such as recurring drought, disturbance, and insect outbreaks, to increase tree mortality resulting in widespread forest die-offs. Breshears and others (2005) described such a case where drought in the early 2000s resulted in mortality of up to 90% among piñon pine (*Pinus edulis*) in piñon/ juniper woodlands across the Southwest. In the northern Rockies, massive outbreaks of mountain pine beetle (*Dendroctonus ponderosae*), a native pest of pines in the western United States and Canada that responds favorably to warm winters, have resulted in greatly increased mortality of lodgepole pine (*Pinus contorta*) and high-elevation five-needle pines (Gibson and others 2008). The combination of mountain pine beetle outbreaks, non-native pathogens, altered fire regimes, and a changing climate may threaten the existence of high-elevation five-needle pine forests (Gibson and others 2008).

EFRs are uniquely positioned to contribute to our understanding of climate change impacts due to their opportunity for research and long-term datasets and their distribution among different ecosystems. Long-term datasets enable EFRs to monitor change over temporal scales that are less feasible for National Forest System units. Please see http://www.fs.fed.us/ccrc/ for a more complete description of climate change impacts in the Rocky Mountain Region.

Management for Adaptation

Millar and others (2007) identified three approaches to managing ecosystems for adaptation to climate change: resistance, resilience, and enabling responses to change. Managing for resistance involves bolstering the capacity of ecosystems to resist changes brought on by climate change. In the case of fire, this could mean devoting resources to the maintenance of historical fire regimes in order to maintain historical species compositions. Other management options are restoring or altering streamflows to maintain riparian areas (Samberg 2011b), irrigating areas affected by drought (Heller and Zavaleta 2009), reintroducing failing or locally extinct populations (Frelich and Reich 2009), locally removing or eradicating invasive species (Millar and others 2007), and intensively managing populations to maintain species distributions (Samberg 2011b).

Ecosystem resilience, meanwhile, is the ability of ecosystems to maintain critical functions and processes in the face of change and disturbance (Samberg 2011a). Strategies for conferring resilience are the introduction of new genotypes to a population for drought resistance or heat tolerance (Heller and Zavaleta 2009), managing fire regimes for the success of desirable (though not necessarily historical) tree species (Samberg 2011b), and maintaining diversity at all levels, from the population to the landscape scale (Samberg 2011b). Management for resilience will typically occur at larger scales and be less intensive than management for resistance (Cole and others 2010).

A third management option in the face of climate change is to assist ecosystems in the switch to states more appropriate for future climatic conditions. In this approach, where particular species compositions or ecosystem structures are no longer the management goals, other goals, such as biodiversity, complexity, or aesthetics, must be defined (Samberg 2011b). Some management actions for facilitating ecosystem transition are assisted migration, creating or maintaining connectivity between protected areas, and maintaining refuges where species may persist while transitions are occurring (Millar and others 2007).

Many proposed adaptation actions, such as introduction of new genotypes, assisted migration, or thinning forests to increase resilience in the face of drought, are largely untried and untested. EFRs could play an important role in investigating the efficacy of proposed adaptation actions. Please see http://www.fs.fed.us/ccrc/ for more information on managing for adaptation in the Rocky Mountain Region.

Local Ecological Knowledge and Oral History

In this project, we examined ecological change in EFRs from the perspective of Forest Service employees who have worked for 10 or more years on particular units. These employees include scientists, technicians, and managers who typically bring both expert, scientific knowledge and field-level, day-to-day experience to the discussion of ecological change. A brief discussion of these different types of knowledge and how they are regarded in the literature helps position this study in the context of previous research.

Local ecological knowledge (LEK) is knowledge held by local people about an ecosystem (Robertson and McGee 2003). Unlike traditional ecological knowledge (TEK), LEK is not the result of thousands of years of cultural engagement with a particular landscape (e.g., the ecological knowledge Australian aboriginals possess about their environment due to tens of thousands of years inhabitation), but it does represent information gained over long time scales by people with a strong connection to ecosystems (Fazey and others 2006). In recent years, there has been increasing recognition of the utility of LEK in ecological research (Bart 2006). LEK has the potential to provide ecological information when there is a lack of more conventional scientific information and to involve local people in natural resource research and management (Robertson and McGee 2003).

Another subset of non-scientific knowledge is experiential knowledge. Experiential knowledge, like LEK, is derived from prolonged interaction with an ecosystem rather than from experiments or formal evidence (Fazey and others 2006). Experiential knowledge was found to be the most commonly used type of knowledge for decision-making among a sample of land managers in Australia (Cook and others. 2009). In Cook and others' 2009 study, as many as 90% of all land management decisions were made using experiential knowledge rather than experiment-based data. Both LEK and experiential knowledge are sometimes referred to as "informal" knowledge and are often differentiated from formal, scientific, or expert knowledge.

As previously mentioned, study participants for this project have both scientific and experiential knowledge of ecological change on EFRs. Some Forest Service staff members have decades-long associations with particular EFRs, and their scientific and experiential knowledge may interact to create a picture of ecosystem change over the history of their tenure with these units.

Research Methods

To better understand the changes that long-term EFR employees have observed over the last few decades and how EFRs can contribute to knowledge of climate change impacts and adaptation, we interviewed current and retired Rocky Mountain Research Station staff members who worked extensively with EFRs. In-depth, semi-structured interviews were conducted utilizing an interview guide to ensure comparability across interviews (see Appendix C).

RMRS employees and retirees with 10 years or more experience working at the EFRs were contacted through phone or email with an interview request. With the exception of one former employee who could not be contacted, all fitting thisdescription were asked for an interview and all agreed to participate for a total of 21 participants.

Participants were given information regarding the basic outline and thematic focus of the interview in order to aid information recall. Of the 21 participants, 8 were retired. There were 3 women and 18 men. Each of the Station's 14 EFRs was represented by at least 2 current or former employees.

Coram Experimental Forest and Tenderfoot Creek Experimental Forest in Montana were the only EFRs not represented by a current employee. The majority of participants were researchers or scientists-in-charge or were otherwise involved in the administration of the EFRs. A small minority of participants were technicians or supervisory foresters. Five participants had worked in their current positions with the EFRs for 10 to 19 years. Eight participants worked between 20 and 29 years at their position with the EFRs, and another 8 participants were with the EFRs for 30 years or more (for a list of participants, see Appendix D).

Three of the participants choose to remain anonymous. The other 18 allowed us to utilize their names in this report. However, many participants covered topics that could be considered sensitive or controversial in nature. These were not anticipated by the research team but are nonetheless highly relevant to the project. Thus, we have removed indentifying information from sections where participants discussed controversial or sensitive topics.

All interviews were conducted in-person by Mason Bradbury. Interviews took place between January and June 2011, lasted between 30 and 90 minutes, and were taped and professionally transcribed. Interviews were coded to identify key themes, and data were analyzed across interviews to discern broader patterns. Analysis provided insights into perceptions of change on individual Experimental Forests and Ranges as well as across the Experimental Forest and Range network. Analysis also examined the future role of these forests in climate change adaptation. Nvivo 7.0 software was used to organize the data.

PROJECT RESULTS

The results of these interviews are conveyed, in large part, through the words of the EFR employees we interviewed to retain the detail and richness of the interview data and to provide evidence for interpretations and conclusions. Direct quotes are italicized to set them apart from the narrative. We refer to the interview participants in this report as either "participants" or "EFR employees" even though some of the interviewees are retired. We report the full range of results and refrain from judgments about the accuracy or relevance of any particular perspective or idea. In other words, if multiple interviewees discussed a particular issue, idea, or project, we deemed it important enough to include in this report.

The following results begin with detailed descriptions of the strengths of the EFRs (how they are uniquely positioned as research sites) and how the institutional context within which they operate offers both challenges and opportunities.

The specific ecological changes observed or documented on the EFRs are then described along with their possible connections to anthropogenic climate change. Finally, participants describe existing and proposed climate change research on the EFRs and the potential role of the EFRs in contributing to knowledge in this arena.

UNDERSTANDING THE UNIQUE STRENGTHS OF EFRs

To better understand the role that EFRs can play in understanding and adapting to climate change, we need to examine the niche that these units fill. Examining the unique strengths of EFRs gives us a sense of their specific contributions to climate change. Some of the strengths discussed by participants were:

- Representativeness. The way in which EFRs represent the different ecotypes that comprise western landscapes.
- Research integrity. How the research focus of EFRs allows the integrity of the science to be maintained over time.
- Long-term datasets. The opportunity to collect data for decades or centuries.
- Experimentation. The ability to host manipulative research that involves treatment of study sites.
- Contributions to management. The application value of EFR research.
- Direct field observations. The way in which long-term field observations inform research design and interpretation of results.

Many of these strengths depend on units that are dedicated to research and allow for long-term study and observation.

Representativeness

One of the strengths or hallmarks of EFRs is their ability to represent broader landscapes or a specific ecotype present in the region. Terrie Jain talked about the Boise Basin Experimental Forest, saying:

> Boise Basin was selected because it reflects the environment that's characteristic of northern Rocky Mountain ponderosa pine forests. And typically, the Experimental Forests are not selected because they're unique, but they're selected because they represent the larger landscape.

In other words, EFRs are not ecologically unique. Rather, they represent ecotypes of interest and offer opportunities to learn more about the broader landscape. Participants also talked about the importance of the landscapes that the EFRs represent landscapes valued for their extent, production of natural resources, and/or embodiment of human-wildland tensions. Stan Kitchen described the importance of the landscapes that the Great Basin Experimental Range represents:

> I believe that there are still great opportunities there, providing opportunities for research in high-elevation treeless landscapes, the subalpine herblands that are kind of the core–the forest areas there are also of interest–together they are representative of a lot of the montane area in the Intermountain West, Utah, some of Colorado, and a little bit of Idaho and Wyoming as well. Those watersheds produce a lot of the water that we use and that's obviously of short supply in the West. So, all of those components would argue for importance.

According to Kitchen, a dominant ecotype at Great Basin Experimental Range is important because of the size of territory it occupies in the Intermountain West and the water it yields. Chuck Troendle described the importance of Fraser Experimental Forest to National Forest Systems as a whole:

> The subalpine/alpine environment at Fraser typifies the central and northern Rocky Mountains. It's very characteristic of a lot of the area throughout the central and northern Rockies. And about 50% of National Forest Systems holdings lie in this particular area, so it's an extremely important representative area.

As Troendle explained, the ecosystem at Fraser Experimental Forest is important to the Forest Service because it occurs on many acres of National Forest lands. Participants also suggested that some EFRs embodied tensions that are widespread in the wildland-urban interface (WUI). Carl Edminster explained in the context of the Manitou Experimental Forest:

> I think Manitou is kind of uniquely positioned because it's right in the Front Range of Colorado. There's actually a subdivision in the middle of it. So it's very much in that Front Range of Colorado tension zone in terms of not only people but environment. And as a result, what happens there and what can be studied there are pretty representative of a lot of the lower elevation forested areas in the Front Range, which pretty much extend from southern Wyoming down into northern New Mexico.

Edminster argued that the fact that Manitou shares similar social and ecological characteristics with the landscape of the Front Range allows its research to be broadly applied.

Research Integrity

Many participants said that EFRs were very important research sites because they provided for research "integrity." In this context, integrity referred to a high degree of control over study sties, protection from management actions that might compromise research, and the ability to continue the research over a long period of time. Bob Denner pointed to this control and protection as something that differentiates EFRs from other National Forest lands:

> Many times, in the early days of research, scientists would go out and install the study on National Forest land, and if you wait long enough, it would have some really long-term, useful information. But, unfortunately, personnel change on the district or the National Forest. And this happens and that happens. And people don't communicate very well. And then next thing you know there's a road built right through a research plot or it's now in the middle of a 100-acre clear-cut. That doesn't happen on an Experimental Forest. We know what goes on where, on each square foot of ground. So if somebody wants to propose a project, they say, "Well, I really like that." And I say, "Well, you better talk to so and so, because they've got something going on on that same spot. Now if your studies are complementary, great. But if there's going to be a conflict, he was there first." So you've got the idea of protection for long-term studies in Experimental Forests.

According to Denner, the control and protection of research is partly due to knowing exactly what has been done in the past on the Experimental Forest and taking that into account when making decisions about future research. He implied that past or on-going studies have priority over proposed projects. Chuck Troendle described what he means by integrity:

> Integrity... the advantage that you have on places like the Fraser Experimental Forest is that they've been administered, maintained, operated in a very consistent manner over a very, very long period of time. And then there's been an absolute dedication to... the utilization

Experimental Forests and Climate Change | 15

of the sites has been dedicated to the research purposes for whatever's going on. Integrity, meaning that over time that has been maintained. Things haven't happened that we don't know about. And in contrast, there's a lot of places where data [are] collected. They quit for some reason, you get more money, they start up. Somebody gets another idea. They quit. They start. [The EFR] is just a long-term, dedicated site.

Troendle described research integrity as coming from an "absolute dedication" to research and a detailed knowledge of what has happened or is currently happening at the site. He argued that EFR research does not vary as much due to changes in funding and new ideas. Michael Ryan also explained how the EFRs' special designation allows them to maintain research integrity:

So Experimental Forests are places with a very special land use designation. And the land use designation is that their primary purpose is research. So Experimental Forests in general allow us to complete research and make sure that if we put in long-term studies, that those long-term studies will, the fate of that will be decided by the scientists as opposed to being based on a land management decision. So it's really a critical piece of the Forest Service Research infrastructure.

The EFR designation is believed to be critical to long-term research integrity, a quality that participants did not perceive to exist everywhere outside of EFRs.

Long-Term Research

Most participants suggested that the ability to perform long-term research was a key strength of EFRs. Similar to many participants, Russ Graham described the long-term research happening at Priest River Experimental Forest, including the dates when projects began:

We have a long continuity of studies. So that gives us a record of long-term forest changes or water change. For example, Priest River, we put in a weather station in the fall of 1911 that is still running in the same place today. We put in a stream gauging dam in approximately 1934. It has been continuously running since 1934. We have tree growth measurements going on to trees since about 1912, 1914. We have plantings that went in on the Experimental

Forest in the fall of 1911 that still exist to this day so we can go out and look at this tree that was planted in the fall of 1911 and show what it looks like today under different conditions.

According to Graham, these long-term studies give scientists at Priest River Experimental Forest a way to see changes happening in forests or hydrological systems. Several participants described long-term research as the defining feature of EFRs rather than just one strength among many. Kelly Elder, describing Fraser Experimental Forest, said "Fraser is an outdoor laboratory. Its value is in its longterm datasets." Elder and the other participants value the long-term records and experiments of the EFRs. EFR employees both value and maintain long-term datasets and research continuity whenever possible, and argued that these long-term datasets are one of the key strengths and contributions of the EFRs.

Science Leadership

Many participants also discussed the leading role that EFRs had played in advancing both basic and applied science. For some participants, the history and historical significance of the research was an important part of the EFRs. Brian Geils said the following about Fort Valley Experimental Forest:

It's a particularly valuable site because it is the first place that we conducted forest research in the United States at a Federal level. There had been other work, of course, done at Biltmore, but this is the first time, and this is where the Federal forest research system was established.

Geils argued that Fort Valley helped establish the Federal government as a major player in forest science. Kelly Elder described the scientific value of early work done at Fraser Experimental Forest:

It's a well-known research entity and the work that has come out of it is well respected worldwide, not just in Colorado and not just in the United States, but worldwide. So some of the fundamental work that has been done in forest hydrology and a few other disciplines was conducted at Fraser.

Another participant described a particular experiment that was "recognized as establishing the relationship between tree canopy removal and increased water yields." Specific EFR units were renowned, even internationally, for early work in specific disciplines. Durant McArthur also discussed the importance of Great Basin Experimental Range to range science and management:

> As far as the Great Basin Experimental Range... it is considered, along with the Jornada Experimental Range, as a cradle of the discipline of range management and range science, and some early pioneering scientists worked there. And many people who became rather prominent in ecology and range science sort of cut their scientific teeth, as it were, at that institution.

Again, a particular EFR was described as critical to the early development of a specific discipline, in this case range science. Bob Denner described the importance of past experiments on Priest River Experimental Forest and the wide range of disciplines to which these experiments have contributed:

> In the realm of forest science, Priest River has a long record of excellence in research and in data collection that's proven to be critically important to both Government scientists and academic researchers in terms of forest development, understanding the mechanisms of fire in the northern Rocky Mountains, fields of genetics in host pest interactions, a long-term weather record that's proven valuable to mensurationists, ecologists, climatologists, hydrologists. You can go on down the list of disciplines that you find within forest science. And there's something there for everyone.

Denner and many others described the broad relevance of EFR research as well as the way EFR studies had contributed to specific management issues. EFRs have played an important role in advancing forest or range science, both in basic and applied setting.

Contributions to Management

Many interview participants emphasized the ways in which EFR research contributed to specific questions and problems in forest and range management. Several EFRs were established specifically to address management issues, including two of the earliest—Fort Valley Experimental Forest and Great Basin Experimental Range. Below, one participant described the impetus for the establishment of Fort Valley:

> At the time this forest [was established], the ponderosa pine forest of northern Arizona was being clear-cut. The loggers had moved in in the 1880s and just decimated a lot of the forest. So, again, these lumbermen, the Riordans, sawmill operators, noticed that the ponderosa pine was not regenerating after the area had been clear-cut. So that's why they contacted Pinchot in 1903 and said "help." That was the main reason for Fort Valley, the silviculture of the ponderosa pine. What it took to get it to regenerate and what affected regeneration. And, of course, they found out clear-cutting was a major factor.

Many EFRs were established to address a specific, applied management question or challenge. Great Basin Experimental Range was established to gain an understanding of the catastrophic floods that affected communities in Sanpete and other central Utah Counties as a result of poor grazing practices. Durant McArthur explained:

> The Great Basin Experimental Range is adjacent to the communities of Sanpete County and was actually established because those communities were being flooded. And the inhabitants petitioned Congress for protection. As a result, first of all the Manti National Forest, which later on became part of the MantiLa Sal National Forest, was established. And then the Great Basin Experimental Range to see what caused these floods and how they could be stopped.

Like Fort Valley, Great Basin Experimental Range was designated with a specific management question in mind, a question of great relevance to nearby human communities. In the case of both Fort Valley Experimental Forest and Great Basin Experimental Range, after the original questions were answered, scientists expanded the research programs to include new questions.

Beyond the impetus for the establishment of some EFRs, participants mentioned that contributions to land management, both public and private, have been strengths of EFRs. Chuck Troendle described the contributions of research at Fraser and Manitou Experimental Forests to Forest Service land management as differentiating them from Experimental Forests in the eastern United States:

> I had come here from the East where there are Experimental Forests that are much more popular, much more well known, like Hubbard Brook or Coweeta. But the bottom line is the research that was being done at that time at those sites was not Forest Service

oriented. It was more ecologically oriented. It was very good research, but it was not being done necessarily in the best interest of the Forest Service. And as a result, a lot of the production that came from those experimental sites, although it was useful ecologically and in the scientific community, it was of very little use to the Forest Service. Whereas, when you come to places like Fraser or Manitou, at that point in time they had not been exposed as much to university or outside control. And, as a result, a lot of the research was still being done with the primary interest being Forest Service application of, or land management application of the information being developed. I thought that was a strength.

Troendle argued that the ability of Fraser and Manitou Experimental Forests to produce information useful to the Forest Service and other land managers was a strength of these EFRs. The leadership role that EFR science has played in advancing specific disciplines was also connected to the application value of such research. In other words, pioneering studies on research topics relevant to management meant that EFRs were a key player in advancing science-based management and helping to answer the questions that managers posed.

Ability to Host Experimental Research

One of the unique strengths that was widely discussed was the ability of EFRs to host experimental (i.e., manipulative) studies. Many participants emphasized the experimental nature of EFRs as one of their distinguishing characteristics. The ability to do manipulative studies was discussed as both a strength of the EFRs and a feature of certain research disciplines. Brian Geils emphasized the importance of the EFRs as places where experimental research can be done:

> We recognize that there are relatively few acres of National Forest land that are reserved and dedicated for doing long-term experimental research. [EFRs are] the place where you should be putting in silvicultural treatments in a scientifically well-designed format that can be maintained over a period of time to see what the consequences of those treatments are.

According to Geils and many other participants, the ability of EFRs to host longterm, experimental research is unique. Carl Edminster also argued that the ability to put in different treatments and observe responses to those treatments is something that differentiates EFRs from other Forest Service land:

> On an Experimental Forest, you've got the ability to do that and then look at the changes that occur over that spectrum of conditions. On the National Forests, a lot of what they are stuck doing is "best management practices," which may be great for certain things but not very good for other things. And because you, in many cases, only have more or less one management action occurring out on the landscape, you're really missing, perhaps, the spectrum of things you could be doing and the spectrum of responses that may occur from those things. So to me, the real advantage of an Experimental Forest is having that living laboratory.

Kelly Elder, Russ Graham, and Terrie Jain also referred to EFRs as "outdoor laboratories" or "living laboratories," as did Wayne Shepperd:

> All of these are outdoor laboratories. If I was talking to a member of Congress, I'd say this is where you do experiments at large scales.

Again, Shepperd emphasized EFRs as places where scientists can do manipulative research. Shepperd also emphasized that EFRs provide a place for experiments at large spatial scales. Russ Graham and Terrie Jain emphasized the value of testing ecological hypotheses through manipulative, silvicultural treatments. According to Jain:

> ...turning around and doing manipulative studies, taking that ecological hypothesis that you developed, implementing it on the ground using silviculture treatments, and then seeing if it actually responds to what you saw.

Graham explained further:

> ...it gives you opportunities to explain those changes [on the Experimental Forest]. Why did they change? Now as a research silviculturist, what separates silviculture from other forest ecology is we go out and manipulate vegetation more than any other scientist. Most sciences are observational and try to explain that phenomena. We not only try to explain that phenomena, we create that

phenomena by cutting trees or disturbing soil or what have you to create or maintain some kind of condition.

Graham explained that the ability to manipulate and recreate ecological phenomena is an important tool for understanding ecological change.

In some cases, EFR employees actively managed the EFRs to ensure that they were representative of surrounding landscapes, utilizing their ability to manipulate to retain the value of the site as typical of a broader area. Wayne Shepperd described the work that was required to create appropriate conditions for experiments on Black Hills Experimental Forest:

> When Bob Alexander and I first inherited the Black Hills Experimental Forest [in the 1980s, it] was an overgrown, overstocked patch of forest in a sea of fairly well-managed forest. And it was totally atypical of any of the surrounding landscape. So from that standpoint, it was sort of meaningless. It wasn't a natural area. It wasn't a wilderness area. It had been disturbed in the past. It just needed a good cleanup, and that's what we gave it. Because we did that, we established conditions that provide experimental opportunities today.

Terrie Jain described the need to actively manage the Black Hills Experimental Forest in the face of bark beetles:

> The Experimental Forests are experiencing the same things that are happening on the larger landscape. We're experiencing diseases, we're experiencing insects. The Black Hills is getting eaten alive by bark beetles. So the thought right now is we cannot just protect them in some state because things were eating them as well. [The scientists in charge] have to actively manage the Experimental Forests.

These excerpts demonstrate the ways that EFRs are regarded, not as untouched, pristine, or untrammeled lands, but rather areas that are actively managed or manipulated for research purposes or to retain their value for future research.

Opportunity for Field Observations

One of the goals of this study was to understand the role of field observations in EFR research. Because interview participants were long-term

EFR employees, some had spent significant time in the field and had the opportunity to observe ecological change through day-to-day observations that were not related to specific experiments/studies. In some cases, field observations took place over several decades at the same EFR. When we asked EFR employees about their field observations, they readily understood what we were asking and consistently described the opportunity to make such observations as a core strength of EFRs and long-term employment with an EFR. That said, the importance that participants attributed to field time and field observations in their own careers varied significantly. Differences seemed to stem, in part, from different responsibilities and research interests, with those participants who did more field-based research valuing field time and field observations more than those doing lab-based or modeling research.

EFR employees whose research did not take them into the field much said that on-the-ground field observations did not play a large role in their understanding of the site or their research. One participant stated "one site is as good as another as far as the science goes."

Some staff members stated that they felt disconnected from some of the EFRs they were associated with because their research was not based on those sites. Other scientists relied on technicians or the scientists on site to provide information about on-site observations to the research. Others discussed the relative benefits of focusing on a range of ecosystems versus emphasizing one ecosystem or site, noting that both approaches were valuable.

That said, the majority of EFR employees valued field observations and readily discussed the benefits of field time. They discussed these benefits in two ways: the benefits of field observations for research and for the management of research. For some EFR employees, field time was critical to effective research management. Chuck Troendle explained the efforts he made to gain first-hand knowledge of the site history at Fraser Experimental Forest:

> The first thing that we did the first summer we were here, we contracted with one of the original scientists at the Rocky Mountain Station to come up and spend a couple weeks with us at Fraser showing us where every historical research site was located, what was done there, what the findings were so that we actually went back in time to try to learn everything there was to know firsthand in the field, on the ground. And from that time on, every study that I was involved in until the day I retired I was personally involved in. One, the design of the experiment, of course, but also the actual site selection, layout, the whole nine yards. Every data collection, I was involved in every part of it. I relied very heavily on technicians, but

Experimental Forests and Climate Change

> in my entire career I never asked a technician to do anything I hadn't done first. And I felt unless you were in the field and walking the watershed, you really wouldn't know what's going on.

Troendle prioritized first-hand, on-the-ground knowledge of project sites and made impressive efforts to obtain that knowledge. The final part of the above quote— "unless you were in the field and walking the watershed, you wouldn't really know what's going on"—indicates the importance that Troendle placed on field observations and field knowledge to the management of research at the EFR. Bob Musselman also illustrated the value of managerial knowledge for doing research:

> I used to go into the field more frequently, but recent management responsibilities have limited my field time. But I can go to a field location and remember what I did there several years ago and how it's changed. So those types of insights, on-the-ground experience of being there I think are pretty valuable to interact with the newer scientists of new projects, even my technician that's only been here 10 years to give him some insight into the changes that have been made; snowcat routes up to the site and that type of thing.

Musselman suggested that "on-the-ground experience" contributes in important ways to the management of research. Linda Joyce explained the importance of time spent in the field by technicians:

> [The technicians' observations] were critical in terms of managing the long-term experiments on site. For example, down at Manitou we had a number of extreme wind events, and one impacted one of the long-term sites. We had a straight line wind event go through. So the observations were critical in terms of managing the long-term experiments and establishing where our short-term experiments might go.

Again, these field observations assisted in the siting of research projects, the longterm management of data-sets, and the overall logistics of project management.

EFR employees also talked about the ways in which field observations contributed more directly to scientific knowledge. According to many EFR employees, field observations deepened their understanding of ecology and ecological change on EFRs, helped them generate research questions or hypotheses, and improved their interpretation of research results. Stan Kitchen described the ongoing nature of his on-the-ground learning at Desert Experimental Range:

> I learned early on that if I was paying attention, I could learn something new with every trip that I took. I would be surprised by something else that I would observe or someone else would point out to me. And you kind of expect that eventually that's going to wear off. It hasn't yet. After 19 years, there are still new things there for me to learn as a person that goes there as often that I do. Some of those things someone might consider trivial or maybe not that important, but I found it to be a fascinating place early on because I love to interact and see what kind of surprises nature can spring on you. And I have not exhausted that yet there.

Wayne Shepperd, in reference to the importance of field observations, said:

> Well, for me, personally, a tremendous role, because as a silviculturist, an applied ecologist, or whatever, just observing things on the ground really goes a long way in helping you formulate hypotheses and questions to guide your research. And I felt that the time I spent on the ground as a technician collecting data tremendously influenced how I approached my research. I was always pretty much a hands-on guy, more of a field researcher than a computer jockey or a modeler or something like that. So in my case, it strongly influenced my ability to do research.

Shepperd specifically stated that field observations help in the formulation of hypotheses and questions that then influence research. Kelly Elder also explained the way in which field observation improved his understanding of ecosystems and helped generate new research questions:

> You can spend years sitting in an office thinking about these problems and the dynamics of them and brilliant people do this all the time and do it very successfully. Not-so-brilliant people can do the same thing with a week of work in the field where you are pounded in the forehead with really interesting research questions and a great understanding of how the systems are working. What you really want is the combination of the two. It is the ability to combine the analytical and academic side along with the field observations that is the most effective. It's another reason that the Experimental Forests are so valuable. Whenever I'm in any kind of stagnation mentally or scientifically, 15 minutes or a couple of hours in the field

and I have more stuff keeping me awake at night and keeping me excited than I can possibly deal with.

According to Elder, field observations inspire new research ideas and help scientists understand how systems work. He also argued that the combination of analytic time and field time helps move the research process along. Ward McCaughey discussed the importance of field time for understanding forest ecology:

> You can read all the books you want about a lodgepole pine forest and how it works, but unless you actually get out there and spend some time in the field and see it and see the variations that can occur with even the lodgepole pine type, that was so beneficial. You can't put dollar signs on it or anything like that. It was extremely important. And I think for any scientist to spend that time in the field is so important to see how the natural world works with your own eyes. And you will understand when you get data back from the field how to interpret what you're seeing.

According to McCaughey, field observations are "extremely important" and particularly critical to data interpretation. Russ Graham agrees, stating:

> I would say that that's probably one of the, maybe even the most valuable part of an informal observation because so many times our formal observation is with a ruler, with an instrument of some kind. And we get to concentrating on the precision of the measurement. And we're not looking at what we're measuring... You might be saying "okay, how long is the root of this tree?" And so you want to make sure that you're measuring it to the nearest centimeter or what have you. But you don't sit back and look at this whole root and say "What is wrong with this root? Is it deformed? Is something wrong with it?" All you're concentrating on is whether it's 22.5 cm or 22.1 cm long. So those informal observations I think are critical to interpreting what your numbers are going to tell you because most of the time all research is around numbers.

Like McCaughey, for Graham, field observations provide context within which to view research results. They help explain the meaning behind the data. Bob Musselman made a similar argument in answering whether or not field observations inform his interpretation of experimental results:

I think yes, although I would like to think research is more objective than subjective. But there are certain things that probably influence how I view my research findings based on what I knew was there 20 years ago. I don't want to say that my research is subjective and it's based on experience but that prior experience [is] obviously involved in my interpretation of the findings. And probably someone that's newer wouldn't have those particular insights on how to interpret the data.

Musselman argued that his long experience at Glacier Lakes Ecosystem Experimental Site (GLEES) conveys specific insights that he would not otherwise have. Carl Edminster, answering the same question, said:

I guess if I had to summarize an answer to that question, I think it's difficult to see things unless you're there witnessing them. You can read about them and be the recipient of reams of data that somebody else gives you. But unless you're actually out there seeing things, experiencing things, I don't know that you have the real depth of perception that you might have if you weren't out there... I think it's really important to be out there experiencing things, not just dealing with the data or reading about them.

Edminster argued that seeing and experiencing things in the field provides researchers a deeper understanding than just dealing with numbers or texts. Michael Ryan made a similar statement:

I think it's all well and good to sit here at the computer and think of ideas but unless you're actually doing a little bit of grappling within yourself or seeing the ecosystem and trying to take measurements in the ecosystem, you don't have the same context for that.

Ryan went on to explain that he is concerned that the opportunity for field time and field observations is being lost.

In short, field observations assist researchers in the siting and managing of research. They also provide important insights into how ecological systems work, ideas for new research projects and specific hypotheses, and contextual knowledge that contributes to improved interpretation of research results.

Summary of Key EFR Strengths

Following, we summarize the strengths of EFRs, as articulated by EFR employees. We revisit these strengths in the context of climate change impacts and adaptation later in this report.

- Temporal scales and integrity

EFRs offer a unique opportunity to conduct long-term research. Long-term research is enabled by research integrity, the level of control and protection that EFR designation gives to research projects.

- Spatial scales and representativeness

EFRs allow for large-scale research, as compared with many other sites/studies. Also, while the scale of EFR research is limited by the size of the EFR, because EFRs represent broader landscapes, the research is applicable over larger scales.

- Manipulation/experimentation

EFRs allow for manipulative, experimental research. Treatments can be conducted and effects can be measured, which might not be possible on other Forest Service lands. This provides opportunities to develop research that examines the impacts of specific management actions, in a relatively controlled environment.

- Relevance/application value

EFR research is relevant to both the science and practice of forest and range management. Research emphases have changed over time, but the EFRs continue to provide information useful to land managers and academics in basic and applied fields.

- Opportunity for field observations to contribute to research

Because EFRs are the sites of many studies over a long period of time, they offer researchers the opportunity to make informal field observations over many years. Field observations contribute to the quality of research by generating ideas for research, refining research questions, and improving interpretation of results.

THE SOCIAL AND INSTITUTIONAL CONTEXT

EFRs exist within a specific institutional context (Forest Service Research, Rocky Mountain Research Station) and within a broader social and political context. Both social and institutional changes influence the ability of EFRs to improve knowledge of climate change impacts and adaptation and to realize the strengths described previously.

Communities and Recreation

We asked all of the interview participants about the relationship between the EFR and local communities. EFR employees described their relationship with local communities in terms of outreach and demonstration, public education, extractive uses, the wildland-urban interface, and recreation.

Public outreach programs, such as field days, open houses, and demonstration days, were mentioned by many participants. Several participants discussed their enjoyment of these outreach programs, but some also mentioned that they are expensive and time-consuming.

Education was pursued for a range of reasons from the need to inform the public on recreational use of the EFRS to dispelling local rumors, such as one about the alleged crash of an alien spacecraft on an EFR in the 1950s. In particular, outreach programs were believed to educate nearby communities about the purposes, rules, and regulations of the EFRs. Chuck Troendle explained his approach to public education:

> While I was there, we worked as much as we could with both the local community and the district. We asked the district to help us educate the public as to what the purposes of the Experimental Forest were, how the rules, the regulations, the uses of the Experimental Forest differed from the uses on the rest of National Forest Systems land. And we tried to engage the public because we found that the public was our best policeman. In other words, they had a vested interest in the Experimental Forest. They didn't want to see it harmed.

Experimental Forests and Climate Change 29

Some EFRs allowed local use of EFR resources for grazing, hunting, fishing, berry picking, and firewood cutting; and in these cases, such uses were an important part of the EFRs relationship with nearby communities. However, in some cases, illegal firewood cutting occurred. Ray Shearer explained how such use can compromise research:

> When you have a road going through the area that's a main road, you have people going up cutting. So we've had one of our plots totally modified and really [it] is not suitable to continue with the other seven.

EFRs in close proximity to rural residential development experienced typical WUI challenges, including conflict over fuel treatment and herbicide application. Kelly Elder explained how management of the Fraser Experimental Forest took into account community needs for fuel reduction:

> We've had great cooperation with the district forest and some community involvement in terms of fuels reduction and fire hazard. We are upstream in terms of fire activity from the town of Fraser, and there are hundreds, if not thousands, of homes down valley from us. If we get a fire in that area, it's going to move right down the valley and here's this huge WUI, the wildland-urban interface, with abundant homes and structures. We worked very closely with the district to do some fuels reductions on the Experimental Forest, which we also incorporated into our research, and to give relief to the district and to the community in terms of some fire mitigation.

Many EFR employees described a noticeable increase in recreational use of the EFRs during the last few decades. They discussed recreation, as a threat to research, a potential topic of study, and a source of change. Kelly Elder described the increase in recreation on Fraser Experimental Forest:

> We have seen exponential increase in recreational use of the Experimental Forest. It used to be very rare to see someone on a weekday in the winter on the Experimental Forest. And even in the summer on a weekday, visitation was spotty, mainly campground use. Now in the summer it's a zoo. On a winter weekday there are, just hazarding a guess, 5 to 20 people cross-country skiing there. And weekends are busy. I would have never dreamed that. It has really, really changed.

One employee noted that increased recreation had come with increased impacts, saying "Recreation, a lot more people in the forest... Just the presence of people, the beginning of roads where there weren't roads, and the trash around the campfires and things like that." Bob Musselman discussed snowmobiling on the Glacier Lakes Ecosystem Experimental Site and the restrictions put in place to prevent interference with monitoring there:

> We've already restricted them from our monitoring area because of our air quality monitoring. We don't want them driving their two-cylinder, smog producing engines around our air monitors.

As Musselman pointed out, certain types of recreation may directly interfere with research results. Other recreational uses mentioned by participants included off-road vehicle (ORV) use (probably the most commonly mentioned, including four-wheelers and motorcycles), cross-country skiing, horse-riding, hiking, camping, and hunting. Michael Ryan described his efforts to protect Manitou Experimental Forest from damage caused by four-wheeler use.

> Lately, something that I've spent a huge amount of time on has been trying to be a little cautious and a little proactive and a little reactive to the recreation pressures and to close roads on Manitou except for, except for administrative use, just to protect the resource.

According to Ryan, soils on Manitou Experimental Forest are very erosive and use from four-wheelers and other ORVs could cause extensive erosion. Russ Graham explained that Boise Basin Experimental Forest also experiences significant recreation pressures:

> The other one, the southeast portion of [Boise Basin Experimental Forest], it's about a 5000-acre unit is just full of ATV [all terrain vehicle] trails and motorcycle trails. So there is a conflict and an opportunity there, very quickly you could see that. There is a conflict of motorcycles, ATVs doing damage to the Experimental Forest. But you have a research opportunity of looking at ATV-motorcycle use from a social, an environmental, or whatever position. So you see there's another case where you have a forest management issue but then you also have a research opportunity.

Graham said he sees the recreational pressures on Boise Basin not only as a threat to research but also an opportunity to research the social and environmental aspects of ORV use. Similarly, Michael Ryan sees an opportunity for research:

> At Manitou, the resource damage from recreation pressure has just really exploded over the last 10 or 15 years. That's just really a big change that I think we can take advantage of that and look at a little bit of the parameters of what the resource damage really is and whatnot.

In short, EFRs operate within a broader social context similar to other Forest Service lands. Relationships with local communities, public perceptions, and recreational use impact EFRs and present both challenges and opportunities.

Institutional Context and Change

While we did not originally have a question about institutional change in our interview guide, initial interviews indicated that EFR employees situated the question of climate change and EFRs in a specific institutional context. Because of the importance of institutional context, we decided to ask each participant about the broader institutional context within which EFRs operate and in what ways institutional changes facilitate and constrain the potential of EFRs to improve our understanding of climate change impacts and adaptation. Many EFR employees discussed changes in research foci and institutional organization. EFR employees were mixed in their evaluation of such changes; some described declining institutional support and diminished research capacity of EFRs, while others described increased support and recognition for EFRs. Due to the potential sensitivity of some of this material, we have removed identifying information from the sections below.

Declining Budgets and Diminishing Capacity

Some participants expressed concerns about declining budgets and decreased research capacity at EFRs. They are concerned about lack of resources to maintain long-term datasets, to fund research and personnel, and to maintain facilities. One EFR employee described a process of shrinking budgets that began in the 1970s. The end result was a cessation of most research activities on the EFR for which he worked.

> And this has come to the point where there's relatively few dollars anymore for research that's funded through the Forest Service entirely because most of it now is going after dollars of most important issue at the time. And what does that mean? It means, in part, that unless you have something that ties in with the program, the National Science Foundation or whatever entity, it may not get funded. So [this EFR] had 10 years of measurements, and it sits there with no continuous research to see what's happening to the stands that were laid out. We've had particular interest by the [nearby National Forest] from time to time on the data collected there, but now there's no further data to provide them. And until there's funding or someone takes interest in it that will go after some funds, it sits there.

Another participant described a similar decline in budgets:

> It's a dismal, tragic decline. The Experimental Forests are falling apart and bleeding to death... Decreased funding, decreased attention, decreased infrastructure, decreased commitment. We have a continual decline in staff, a decline in funding, a decline in infrastructure, a decline in interest except at the grass-roots level. We're talking about the jewels of the Rocky Mountain Research Station. [They] are important for Forest Service research across the country but there's no support. ...We lose studies. I'm closing down one of [my experiments]. I can't afford to fix it, and I can't afford to keep it running, and I can't do it myself anymore. So we're starting to lose data. And, again, the value of these Experimental Forests is, in fact, the long-term data. And if we're cutting programs and cutting data, then pretty soon, at some point, you'll reach a point where you might as well turn it back over to multiple use and get rid of research. We're not there yet, thankfully, but things aren't looking good. Science suffers.

Again, this participant pointed out that declining budgets compromise the ability of EFRs to maintain long-term studies. Another participant described the way he used Station funds to supplement external research grants in the early days of his career. He would use Forest Service dollars to maintain his lab, buy instrumentation, and pay a research assistant. Now, he argued that Forest Service funds only support buildings and salary, with little left for research expenses. He views this as similar to universities, but stated that

Forest Service researchers do not have the same freedom as academics. Another participant argued that more and more Station funding was being allocated to overhead with a lower proportion available for discretionary research support, as compared with the past. Several participants suggested that the efficiency of the station was declining due to the way that budgets were being allocated. Another participant described the declining number of scientists at the Station:

> There's fewer of us and not any less work... In the Rocky Mountain Station alone [previous to the 1997 merger of the Intermountain and Rocky Mountain Forest and Range Research Stations], we had 100 scientists. Now through the whole Rocky Mountain [RMRS post-merger], we have 91... So administratively, that's probably the biggest shift. And then also we have to look for more funding to sustain ourselves. Historically, nobody ever had to put in proposals. They did their research. They could focus totally on their research scenario and answer those questions, versus now we got to go find money... We've got to do a bunch of other things, and that takes us away from focusing purely on the science end of the research. So we're juggling more balls in the air.

In his experience, a declining budget has meant fewer scientists and more reliance on external funding. One participant described the current strategy at RMRS of not rehiring for research technician positions at the EFRs after they have retired:

> We're not hiring any permanent technicians or professionals at the station. We do it through post-docs, we do it through temporaries, some term positions we fill that role. And that's a different philosophy. It's more of an academic or university philosophy. [For] much of our research program, that might be very appropriate. But my contention is that for these long-term places, like all the Experimental Forests and Ranges, we need people with corporate memory, intimate familiarity with the locations. In my mind, at least, that means a permanent employee.

If field observations contribute to the quality of research (as previously described), then this participant's argument that EFRs "need people with corporate memory, intimate familiarity with the locations" is particularly important. Also, several of the participants we quoted connected declining budgets to a decreased ability to continue collecting data for the long term. This is particularly important given the fact that many people discussed long-term research as one of the key strengths of EFRs. In some cases, EFR employees were so invested in particular studies that they found ways to continue the research despite lack of funding, saying that they "continued them in spite of the station not because of the station."

EFR employees also expressed concern about the need to effectively manage and archive long-term data. They discussed the need to digitize old, handwritten records and a need to digitally map current and past research on the EFRs. They articulated concerns about data being lost due to outdated modes of archiving information. Some mentioned a need to find appropriate storage space for the original hard copies. One participant suggested that only 1% of the funding for EFRs was dedicated to data management, as compared with long-term ecological research sites, which he argued spend 50 to 60% of the budget on data management. Without the funding to maintain and manage long-term datasets, EFR employees feared that the value, accessibility, and application of long-term research would decline. EFR employees also discussed the expense of maintenance at EFRs with more extensive facilities and the impact of declining budgets on their ability to maintain on-site facilities.

Overall, participants expressed concerns that declining budgets meant increased reliance on external funding (discussed in more detail below), decreased ability to conduct research, fewer scientists, inability to maintain long-term studies, inability to effectively manage datasets, and less maintenance of facilities.

Increasing Support and Recognition

The feeling that the EFRs are not adequately supported was not universal, as some participants talked about a recent increase in the level of support and recognition for the EFRs coming from the RMRS and higher in the Forest Service. This trend was characterized as "heading in the right direction." In the following quote, one participant discussed a change toward better funding of the EFRs in recent years:

> The Station, to its credit, has provided a budget for the Experimental Forests. Before about three or four years ago, it didn't. Individual projects had to keep it going, and there often was not enough money. So there's been more of a commitment from the Station in the past three, four years to Experimental Forests to provide a budget.

Another participant noticed a change in the level of interest among RMRS staff:

> The last decade at least, it seems to me that the Rocky Mountain Research people are more interested in what [this EFR] is and can be... I have seen a real interest in preserving both the site and these records.

In several cases, participants acknowledged the increase in recognition given to the EFRs but still noted issues with funding. According to another participant:

> There's got to be a recognition. And I believe there is, that there is very high value on Experimental Forests within Rocky Mountain Research, within the leadership. The struggle is how do you slice up the pie to maintain all these [Experimental Forests], and are we going to reach a point where one or more Experimental Forests may have to be de-established because we can't afford them anymore.

He acknowledged support from RMRS leadership and the challenges of allocating a limited budget. One participant explained that there is a lot of support for EFRs in higher administration, but that there are leaders who feel that EFRs are unnecessary:

> A lot of our program managers appreciate Experimental Forests and understand them, but there are a few that don't. And they figure it's a money sink. In a way it is, but [the question is] whether or not that's a valuable money sink and a long-term investment in the future. I think it is. My view is the longer the record the more valuable it becomes. But it's expensive to maintain those records, and if you have to cut, you have to cut somewhere. And some people think, well, you don't need Experimental Forests. It's an easy way to cut.

Like many others, this participant emphasized the importance of investing in long-term data collection.

Changing Research Foci

EFR employees described several reasons for changing research foci on the EFRs. In some cases, the research questions that had driven establishment of the EFR had been answered. New issues emerged in forest management and on the EFRs, such as fire and insect outbreaks, resulting in new research questions. In other cases, Forest Service and public interests in non-timber resources and values led to a diversification of research on EFRs. EFR employees also talked about the ways in which funding shifts drove research foci on EFRs.

Many participants said that ecological change on the EFRs influenced research. Bob Musselman weighed in on whether ecological change has affected research at GLEES, saying "Yeah, definitely the focus of research. We wouldn't be studying beetle kill up there if the trees weren't dying." Dan Neary discussed the influence of big fires on the research focus at Sierra Ancha and Fort Valley Experimental Forests:

> The fire definitely has. It's been driving things. Sierra Ancha... Fort Valley was the same way because there was a lot of restoration started because they realized that they had to reduce fuels. And so that was set off by a series of fires in '96. So they had a really big impact on work that was going on in the forest since.

One of the consequences of shrinking Forest Service budgets for research that participants described was a shift toward more reliance on external funding, or "soft money," as discussed briefly above. Kelly Elder described the change toward externally funded research, characterizing it as something that took place as he began working with the EFRs:

> I came from the university and worked using soft money because that's the way university research works. I stepped into this job expecting and being trained in raising soft money and doing soft money research. And it was a good thing, because I came in at about the transition where the corporate dollars for research really dwindled. And so the change you see at the Experimental Forest level is now the research is largely guided by the ability to bring in soft money. As soon as you start bringing in soft money, you change the questions that you ask, because you are responding to RFPs [Requests for Proposals] or funding requests from other agencies who have interest in specific research. We have had to become creative in maintaining Forest Service research due to reduced funding and long-term studies have been particularly difficult to maintain. What we do now is compete for funding that serves the needs of both the Forest Service and other funding entities.

Experimental Forests and Climate Change

Elder and others believe that external funding is changing the kinds of research questions and topics pursued on the EFRs, and that the resulting research might not be as relevant to Forest Service managers, as compared with research funded by Station dollars. Carl Edminster made a similar statement about the shift toward shorter-term research:

> I think that there has been a move in many cases away from the longer-term research to the shorter term, whatever the emerging issue of the day happens to be.

The concern here is the external funding might not support long-term research, partly because it responds to trends and current interests, as opposed to longer-term questions. Overall, external funding was widely believed to have shifted the focus of research on the EFRs, at least to some extent. This shift toward external funding was regarded as negative by some participants and as more benign by others.

Participants also described the way in which different fields and disciplines have been emphasized or prioritized at different moments in the history of the EFRs. One participant suggested that the Station had eliminated research in key areas, such as fire and range and watershed management. This participant also articulated the challenge of predicting what research topics will be most important in the future and thus which long-term research to prioritize. Russ Graham claimed that the mindset or frame in which forests are viewed has changed over the course of his career:

> In the mid-1970s when I started as a Research Scientist, we were heavily into timber production. Timber production was the main goal of the Forest Service. So, all of the projects and all the research from about 1975 through 1983 were aimed at timber production. Everything was fertilization, making trees grow faster, straighter, minimizing any kind of competition so that your crop trees would grow. But then about 1983 or so, we started looking into other aspects of the forests. [We] started looking at the forest floor, the lower vegetation, even looking at the root systems of trees, the ectomycorrhizae in root systems, that type of thing... So at first, it was always just pretty much looking at the tree, just the tree as a product. Then by the 1980s, we started looking at ectomycorrhizae. We looked at soil processes. We looked at productivity. So while the forest is changing, also the expectations from society were changing and the expectations of what research we were doing were changing. So when you ask me how have the forests changed, it changed not

only that trees grew but also what kind of experiments we were doing. And so when you look through that entire history of Priest River, like I say, from 1911, nearly 100 years now, the forests were changing but also the research and the expectations and the values of the research was changing. But always it comes back to that that forest was the place that still offered those opportunities as changing research.

Graham described the expansion of EFR research from a focus on timber production to a range of studies on different aspects of forest ecology and how this shift was prompted by ecological changes and by changing societal expectations. Terrie Jain described a similar shift from a perspective within which timber was the top priority to a range of research interests:

In the early part of the century, we were really into timber production. We were growing trees for products, per se. Well today it's a whole different game. We're talking about resilience and restoration and integrated objectives from wildlife to all of those. So those are different than timber production. These are much more complex, much more difficult. And then there's climate change adding its whole little tidbit to it... The other thing that's changed is we're looking at larger landscapes now versus individual half-acre plots' growth.

Jain argued that movement to a focus on different aspects of forest ecology and management, including resilience and restoration and large landscapes, was more complex and difficult.

Wayne Shepperd also described similar changes in the research program at Manitou:

...came a shift in funding more to timber management research and multifunctional research... We essentially dropped the range research at Manitou and started more on hydrologic research, more basic research and forestry and forest ecology research. And then that shifted again during the Hayman Fire because we got money to do fire research. So we went from timber management to fire recovery and then to urban interface fuels treatment research, which is still being done down there today.

Shifting research foci was attributed to the interests of external funders, public values and perceptions, and management challenges on National

Forests. Interestingly, Wayne Shepperd suggested that the long-term datasets from EFRs present opportunities for new short-term research:

> There's certainly been a shift from long-term research to short-term research, but that doesn't mean that we need to dump the Experimental Forests. If anything, that makes them more important because the long-term records and the long-term activities that have been documented there provide a tremendous tool, leverage, if you will, to do new types of research. And we see examples of that in [the Manitou, Fraser, and Black Hills Experimental Forests] right now.

According to Shepperd, long-term data can be leveraged in the development of new research directions on EFRs.

Relationship with National Forest Systems Management

The relationship between EFR employees and National Forest staff also presented both opportunities and challenges. The EFRs are located on land owned by National Forest Systems, with the exception of Desert Experimental Range, which is owned by the Rocky Mountain Research Station. Thus, most EFRs rely on National Forests for critical management activities, such as road maintenance and research-oriented timber sales. Bob Denner discussed the range of work on the EFRs provided by the local ranger district:

> The activities that we do on the forest are restricted to research only. So road maintenance, timber sales, contract compliance, post-harvest, silvicultural activities like thinning and planting, law enforcement, fire protection are all provided by the local ranger districts. We spend a lot of time maintaining good working relationships with the local district personnel at each of those sites.

Employees within more management-oriented disciplines, such as silviculturalists, relied on the National Forests to do the treatments, such as timber harvests, required for research. Terrie Jain described the importance of maintaining good relationships with the local NFS district for doing her research:

> [The relationship with the NFS district] provides opportunities... The 650-acre study I put in, I couldn't have done it without them. They did all the layout. They paid for the NEPA [required assessments of environmental impact under the National

Environmental Protection Act]. They paid for the paint. I hired somebody to go and help mark... as far as the research goes. They are going to do the burning. They did all the thinning in the two weird designs I wanted. But it was all their money and all their effort and all their contracting to do it. I wouldn't have been able to do a single thing without that. So it's huge. Huge.

Another participant mentioned that NFS management helps with the protection of monitoring installments as well. Such protection included law enforcement and the removal of hazard trees from the vicinity of monitoring sites. Good working relationships are particularly important because, according to one participant, there is currently no document formally establishing the priority of research over other uses.

There is not any formal document that says that research takes priority over all other uses. Hopefully, we're going to correct that in the next round of Forest Service manual revision. So it really depends on the working relationship between the Experimental Forest and the district as to how well research is protected.

Thus, the relationship between the National Forest and the EFR is particularly important for maintenance of forest infrastructure and for implementation of experimental treatments/manipulations.

Summary of Social and Institutional Changes

Following, we summarize some of the key social and institutional changes that participants described. While these were often articulated as challenges, many participants also see opportunities in these changes. In particular, there were mixed views on the extent to which the station and agency supported the EFRs and on the impact of external funding. Both social and institutional changes are relevant to the ability of the EFRs to contribute to research on climate change impacts and adaptation as they both enable and constrain such research.

- Increasing recreational use

Increasing recreational use of EFRs requires management of visitors and restrictions to ensure that data collection/research sites are not compromised.

Recreational use also presented opportunities to study recreational impacts on key resources.

- Declining budgets

Declining budgets were of widespread concern. Declining budgets are believed to result in an overall inability to maintain staff, continue research, effectively manage data, and, in some cases, maintain facilities. The biggest concern with declining budgets is the inability to continue long-term studies. A few participants suggested there was increasing support and recognition of EFRs.

- Increasing reliance on external funding

An increased reliance on external funding sources (non-Forest Service) due, in part, to declining budgets is seen as shifting research priorities and, in particular, limiting the focus on long-term studies.

- Expanding research focus

The expansion of research on the EFRs, beyond timber production or range science, is seen as both a challenge and an opportunity. A wider range of questions are being asked, but research topics are increasingly complex and occur at larger spatial scales.

- Maintaining relationships with National Forests

The need to maintain good working relationships with National Forests to ensure the maintenance of forest infrastructure (e.g., roads) and support for treatments (e.g., thinning or burning) is of continued importance to EFRs.

THE ECOLOGICAL CONTEXT

Participants were asked to describe the ecological changes they observed on EFRs and to discuss what changes might be attributed to anthropogenic climate change. EFR employees drew on both formal studies and informal, on-the-ground field observations to describe these changes. They focused on both "natural" change, such as succession and weather variability, as well as

Forest Change

Many EFR employees described the ways EFR forests had changed over time with respect to structure, function, and species composition. Several participants stated that the EFRs looked very different now than at the start of their careers. Wyman Schmidt explained that most of the ecological change he saw on Coram Experimental Forest was due to succession:

> I think most of it is just purely natural succession processes... As the trees get older, they provide more shade and you start getting shade-tolerant species coming in and they, in turn, they affect the understory and how the understory is available for wildlife.

Wayne Shepperd described the growth he has seen at Black Hills and Fraser Experimental Forests since the beginning of his career:

> These are dynamic systems. I was literally able to see the forest change in density and growth over time during my career, especially at the Black Hills. Trees that were sapling size when I first started and first saw that forest are now saw logs. And the regeneration that started is just amazing. So I've seen those changes. I've seen similar changes, but on a much slower scale, at Fraser on some of our regeneration studies. We've got saplings that you could go up there and you could find a tag on the bottom of them that we found that tree the week that it germinated and it's now a sapling size tree. And it's taken it 45 years to get there. So it gives you an appreciation of the environment that those forests are trying to exist in.

Russ Graham discussed small-scale disturbance and the processes of growth and succession:

> When we look at forest changes, I would suggest, trees grow. And one of the most august things at Priest River and Deception Creek, one of the biggest forest changes, was the mechanical disturbance of weather. About every five to six years on one of those forests there is a wind event, a weather event, a snow event that would change some of the character of the forest. A lot of people I don't think recognized how those small disturbances interacted with

Experimental Forests and Climate Change

the forest growth and development. Also, then you would start seeing how different trees would respond to those disturbances. Western white pine, for example, is a very fine needled pine, very loose. Meanwhile, grand fir is a much heavier needled tree that captures snow. For example, then you would have this big snow event where snow would be falling out of a tree. The white pine would shed that snow and the grand fir would get broken.

In some cases, forests change was attributed to insects, disease, and invasive plants. The recent mountain pine beetle outbreak was the most commonly mentioned, but white pine blister rust (*Cronaritium ribicola*), larch casebearer (*Coleophora laricella*), and spruce budworm (*Choristoneura* species) were also discussed. Kelly Elder told of the impact of the mountain pine beetle on Fraser Experimental Forest:

> The biggest change I've seen at Fraser is that everybody notices, even the people who live in the cities and drive through there for the first time, is the impact of the mountain pine beetle. We've seen major mortality in the lodgepole forests at Fraser with up to 90% of the living lodgepole decimated. We are quickly seeing a response to that impact in terms of vegetation with radical growth in the understory vegetation that's already established there. We think we'll see a shift in species composition. Certainly, we're seeing more subalpine fir coming in and that's traditionally been a mixed lodgepole pine, Engelmann spruce, and subalpine fir forest. It has been the lodgepole that has been pounded, but they are coming back too. We are also seeing aspen taking off.

Ray Shearer described a western spruce budworm outbreak on Coram Experimental Forest:

> Well, you had the periodic western spruce budworm that defoliated all the subalpine fir and, to a lesser degree, the Douglas fir. Very little on spruce and none on western larch that could be discerned. But, nevertheless, that peaked and then waned so we had a period of maybe 5 to 10 years where the budworm was evident. And as far as I know, it hasn't been evident for 20 years.

Terrie Jain described the cascading effects of white pine blister rust on the ecology of Deception Creek Experimental Forest:

Well, there are changes going on. There are two things that interact. One, in Deception Creek, for example, we had the introduction of blister rust, which really knocked out the white pine. One of the other species that took over that niche was grand fir. What we historically had was a mixture of tree species; now, we have more of a dominant one species system. And we're seeing at Deception Creek things like tussock moth and other things coming in and taking a number on those grand fir. Now I can't say that it's climate change or is it the food is so available they're just taking it in, because grand fir isn't really resilient to a lot of diseases and insects and endemic insects. So there are two things that happen. One, we had a species shift. And two, now we have a species that is not as resilient to the endemic disturbances that functioned.

Desert Experimental Range experienced the introduction of several weedy exotics. Stan Kitchen described these introductions as the primary agent of vegetation change Desert Experimental Range has experienced:

The biggest change that I think that we have seen has been the expansion of some introduced weeds that are becoming better established, in some instances replacing the native vegetation.

Wyman Schmidt described the initial infestation and later control of the larch case-bearer, an introduced pest of western larch at Coram Experimental Forest.

The advent, the introduction of the larch casebearer, for example. It was a big intrusion into the system, but because of some of the research that was involved with that and we were able to get on top of much of that problem with introduced parasites and predators. So that was relatively short-lived, you know, in terms of forestry.

Fire was another agent of change discussed by EFR employees. In particular, the increase in higher intensity, "catastrophic" fires was noted by several participants as having significant effects on the ecology of certain EFRs. Dan Neary described the advent of large, high intensity wildfires on Sierra Ancha and Fort Valley Experimental Forests:

The big one has been fire, wildfire. That really wasn't there before. Sierra Ancha had a fire back in the '50s, but it wasn't really big. And throughout the Southwest, a lot of these fires, now they're bigger and badder. And this was a big, bad one. Fort Valley's had a couple minor ones that it didn't really have in the past.

Similarly, Wayne Shepperd described the occurrence of such fires as a change he has seen over his career on Manitou and Fraser Experimental Forests:

> The other change I think that's profound in the Front Range is the occurrence of the fires. They're much larger now than they used to be... And we had tremendous numbers of fires. The Hayman Fire was the largest fire in Colorado history, which burned in the forest. So that was something different than our historic records showed existed in the past.

Participants see EFR forests and ranges as dynamic ecosystems, and they readily described many different changes to the EFRs in which they worked. EFRs had experienced shifts in species composition due to the process of succession; insect and disease outbreaks (some which had come and gone, others on the rise); changes due to disturbance events such as wind storms, heavy snows, or fires; changes to the disturbance events, such as the increase in higher intensity fires; and invasion by non-native species, primarily invasive plants.

Climate Change

EFR employees also described a suite of changes potentially related to climate change, including many of the changes previously described. Warmer temperatures and changing precipitation patterns were widely believed to impact species composition, insects and disease, and wildlife. Dan Neary explained that the combination of climate change, drought, and fire has caused conversions on the Arizona Experimental Forests:

> There's been thousands of acres in some places with type conversion. It's often going towards more drier species. In other words, you're going from ponderosa to chaparral, for instance, or grassland because you've got the zones as you go up in altitude here. The climate change, the drying out tends to, shoves things lower so you have, in some other places you've had over the past 50, 70 years piñon-juniper's expanded because it's been wet. Well, now being dry and hot and getting some burning, it's retreating again to higher

altitudes. That's real visible. It's not always necessarily visible on the Experimental Forest, but as I said, we've had fires like the western most section of Fort Valley went up in a wildfire and had been used for some research.

Another participant described changes in growth form in Engelmann spruce (*Picea englelmannii*) on Fraser Experimental Forest as a result of milder winters:

> Well, the principal change I saw, there's an area that we call Alexander's Knoll that is pretty much on the edge of the alpine going from the subalpine to the alpine. And there was a lot, on this particular knoll, there was a lot of Engelmann spruce krummholz in the early '80s. And during my time there, I have seen that krummholz start sending up main shoots and look like it's turning into a tree instead of just krummholz. And I associate that with less rugged weather conditions, milder weather conditions, milder winters.

Stan Kitchen explained that increased mortality in Engelmann spruce on Great Basin Experimental Range is likely tied to milder winters:

> In the Great Basin Experimental Range, probably the most obvious change if someone had been up there 20 years ago and went up there now is the change that has happened in much of the West, and that is we've lost most of our Engelmann spruce. And they're standing in red. It's a very obvious change that's taken place. There's reason to believe that that is linked to climate change and with milder winters. We don't have real clear temperature data over the last two years... But winters in general have been milder. And that's believed to be tied to that die off.

Wayne Shepperd explained why he believes the incidence of sudden aspen decline may be attributable to climate change:

> In the case of sudden aspen decline, the fact that Jerry Rehfeldt's model showed that the distribution of or the area in which aspen will be capable of growing, the climate in that area will no longer sustain aspen in the end of the century. The areas that his model projects will no longer be able to sustain aspen are the very same areas that we've seen a predominance of sudden aspen decline. There's almost a one-to-one agreement there. I don't think that's a coincidence.

Bob Musselman explained that climate change, in association with other factors, seems to be driving spruce and fir beetle infestations at GLEES.

> Climate change is driving the beetle issue as well. The direct effect is the beetles are killing the trees. But the indirect effect is that climate change is causing the beetles to explode. Well, not all climate. The stand is failing. It's an old-growth stand and it's weaker. The drought stressed it. So those are involved. Drought stress, old stand, and it's more susceptible to beetle kill. I think the climate change is the driver. And we're trying to document what those climate changes are, and we're seeing some changes in temperatures some months of the year in just 20 years. There are changes in snowmelt rates and initiation of snowmelt, that type thing.

Another participant explained that the mountain pine beetle infestation is similarly tied to moderating winter temperatures along with drought:

> Well, obviously the mountain pine beetle. And that all pretty much kicked off as a result of the drought in 2003. It stressed the trees and made them susceptible. And then we haven't had the bitterly cold winters, 40 below temperatures that you need to knock them back like you used to have in the '80s and there usually was a week during the winter when it would go down to 20 below 0 every night in Fort Collins. And I presume that would correspond with temperatures going down to 40 below in the Fraser Experimental Forest area. And we haven't seen that since the early '90s. We had a little bit of sub-zero weather this year, but nothing like we had in the good old days.

Michael Ryan explained changes over the last three decades in the fledging dates of the flammulated owl (*Otus flammeolus*):

> He's got a really strong signal that shows that over the 30 years he's been studying it [the owls] have fledged, left the nest 10 days earlier on average... And that's related to the climate there getting warmer.

In short, many different changes were attributed to climate change, drawing on both formal studies and informal observation. These changes included the decline of certain species (e.g., aspen or Engelmann spruce), conversion of some areas to new habitat types (e.g., conversion of ponderosa pine forests to chaparral), changes in forest structure (e.g., from krummholz to trees), increase in native and non-native pathogens (e.g., mountain pine beetle and spruce budworm), and shifts in the timing of wildlife activities (e.g., the fledging of owls).

Natural Variability

Many EFR employees discussed both anthropogenic climate change and natural variability in climate and weather as sources of change on the EFRs. However, it should be noted that changes attributed to climate change by some participants were described as part of natural variability by other participants. Bob Denner explained that climatic cycles give rise to pest infestations on Priest River Experimental Forest:

> But again it's cyclic... We've seen outbreaks of insects, but that was a direct result of a weather event that created ideal conditions for the endemic population to explode. Once that subsided, everything went back to some semblance of normalcy.

Chuck Troendle explained the cyclical nature of precipitation in Fraser Experimental Forest:

> There are cyclical patterns in our data. You asked about the longterm data. We see about a 12-year cycle in our data. In other words, you have a wet cycle and then you have a dry cycle. And that [cycle] is about 12 years... And that varies around the country so that it just depends on where you're at.

Ray Shearer claimed that variation in precipitation has been one of the drivers of ecological change on Coram Experimental Forest:

> Well, precipitation is the key factor. Some summers we would have relatively frequent showers that would prolong the life of new seedlings. And other summers by the 4th of July most of them would be gone. So there are these kinds of variations.

Stan Kitchen made a similar argument about the paramount importance of precipitation, linking precipitation cycles to changes in wildlife distributions on the Desert Experimental Range:

The wildlife portion cycles very much. Rabbit, jackrabbit population cycle up and down through the years. And the predators cycle with the primary consumers. And I've seen that variability from a lot of rabbits to few and a lot of rodents to few and the hawks and owls and everything that prey on those. Kit fox as well that we have had research, though I haven't had my own personal research, but we have had others that have, have seen that variability. But it essentially has tracked the overriding pattern of that annual to oftentimes multi-year pattern of variability in precipitation, as that would be the driving force.

Russ Graham argued that year-to-year variability in weather at Priest River Experimental Forest has been more noticeable than directional change in climate:

In my time, at Priest River especially and the long-term weather records, [I've seen] how unaverage I guess the weather is really. And there's from one year to the next you can't really say that it's going to be wet or it's going to be dry or what have you. So anyway, when you talk about changes, I would say not as changes as much as the variation of weather that you can see over time in Priest River, for example.

Keep in mind that many participants discussed both anthropogenic climate change and natural variability, and that participants who described processes of natural change did not necessarily dismiss anthropogenic climate change as an important driver of change.

Summary of Ecological Changes

Following, we summarize the key ecological changes that participants discussed. EFR employees drew on formal research results and informal observations to describe these changes.

- Dynamic ecosystems

EFR ecosystems were characterized as dynamic systems in which many different kinds of change were expected, observed, and documented.

- A variety of ecosystem changes

Changes to EFR ecosystems included shifts in species composition due to successional processes and invasion by non-native species, the influence of the growth and decline of various insects and diseases, the impacts of disturbance processes (e.g., wind, fire, and heavy snows), and changes to the disturbance processes (e.g., larger or more intense fires).

- Anthropogenic climate change resulting in species decline, habitat conversion, increases in pathogens, and timing changes

EFR employees attributed many ecological changes to anthropogenic climate change, including the decline of certain species (e.g., aspen or Engelmann spruce), conversion of some areas to new habitat types (e.g., conversion of ponderosa pine forests to chaparral), changes in forest structure (e.g., from krummholz to trees), increase in native and non-native pathogens (e.g., mountain pine beetle and spruce budworm), and shifts in the timing of wildlife activities (e.g., the fledging of owls).

- Natural cycles and variability are important

EFR employees also acknowledged the important role of precipitation cycles and the natural variability of weather patterns. They recognized the ways in which precipitation cycles impacted ecosystem features such as wildlife populations. Many participants acknowledged both the influence of natural cycles and the effects of anthropogenic climate change.

EFRs and Climate Change Research

To assess the capacity of EFRs to contribute to climate change, we asked participants to describe existing or proposed climate change research and to talk about the challenges and opportunities of future climate change research on EFRs.

Monitoring and Modeling

Many existing monitoring and modeling studies associated with EFRs were contributing to knowledge of climate change. Participants described weather station data from the EFRs dating back to their inceptions, as well atmospheric monitoring as part of broader networks such as National Ecological Observatory Network, National Center for Atmospheric Research, and the AmeriFlux system. Several EFRs were appropriate for this type of monitoring because of their relative isolation and the control over the monitoring installations that EFR designation offered. Long-term data on temperature and precipitation were not only important because they contributed to knowledge of regional and global climate trends. Such data could also be combined with on-the-ground ecological measures to help researchers understand the relationship between changes in climate and changes to EFR ecosystems.

EFR employees also discussed the ways in which existing monitoring and modeling might help us understand how certain species were responding to changing conditions. For example, Bob Musselman talked about vegetation monitoring of beetle kill plots at GLEES, and Carl Edminster discussed work monitoring species' responses to drought. Russ Graham mentioned a modeling study by a former RMRS employee looking at shifts in seed viability zones in the context of climate change. The model was constructed with data coming primarily from Experimental Forests. Chuck Troendle discussed a modeling study on the impacts of climate change on subalpine hydrology. Kelly Elder described hydrological monitoring being conducted at Fraser Experimental Forest that could be used in adaptive management plans.

Research on Climate Change Impacts

EFR employees discussed a range of research projects or proposals focused on climate change impacts, including studies of changes in snowpack, vegetation composition, hydrology, wildlife distributions, and soil processes. The list of projects in this section is not exhaustive but rather provides a sense of the range of existing and proposed climate change studies on EFRs.

Bob Musselman mentioned research at GLEES looking at changes in snowpack and vegetation composition over the past 20 years. Fraser Experimental Forest is part of a collaborative project looking at the ecology of the gray jay (*Perisoreus canadensis*). Brian Geils mentioned a proposal for a then and now comparison at historical weather station sites on Fort Valley Experimental Forest. Dan Neary conveyed that the Arizona Experimental Forests have been asked to join a study of decomposition under different carbon dioxide concentrations.

Participants also described opportunities to utilize long-term datasets that were not originally designed to examine climate change but were uniquely positioned to answer climate change related questions due to the specific data collected and the long timeframe. Kelly Elder explained that much of Fraser Experimental Forest's research in hydrology and biogeochemistry should show effects of climate change. Linda Joyce and Michael Ryan described a collaborative project at Manitou Experimental Forest with the Colorado College studying change in the flammulated owl. Michael Ryan discussed how the project can contribute to climate change knowledge because of the long timeframe and the nature of the study:

> Because of the long-term nature of it, we have a really unique long-term study there that's run by Brian Linkhart out of Colorado College who has been studying this tiny little migrating owl called the flammulated owl. It's about 6 inches tall maybe at the biggest. And they go, we think, down to Mexico for the winter and then come back here. And so because Brian has been studying that for 30 years, we know an awful lot about this bird, their fidelity to different trees when they come back to different mates... One of the really unique things about that particular dataset that unfortunately isn't published yet but is that he's got a really strong signal that shows that over the 30 years he's been studying it they have, they have fledged, left the nest like 10 days earlier on average because... And that's related to the climate there getting warmer.

In other cases, new studies had been designed to answer emerging climate change questions. Stan Kitchen described research at Great Basin Experimental Range looking at responses to manipulations in precipitation:

> At the Great Basin Experimental Range currently there's a study that involves... rainout shelters, so it's redistribution of precipitation so that, assuming that with climate change there could be some changes in if not the total precipitation, the timing of precipitation. So there's some rainout shelters to redistribute through time the precipitation. It's been through two years of treatment at this point. And they're looking at soil processes and vegetation processes across an elevational gradient. So shrublands, forest, forest openings, and then up in the subalpine area, nonforested as well.

Thus, EFR staff were taking advantage of existing, long-term studies to gain new insights on climate change impacts, and designing new studies to specifically look at climate change.

Research on Climate Change Adaptation

We also asked participants about research looking at the outcomes and effectiveness of climate change adaptation actions (i.e., management actions taken in response to climate change impacts). Several participants discussed the ways that manipulative experiments could improve knowledge of adaptation, with the most frequently mentioned being common garden studies. Bob Denner and Russ Graham described common garden studies at Priest River and Deception Creek Experimental Forests, with another common garden study set to be implemented at Priest River Experimental Forest in 2012 in collaboration with the Canadian Ministry of Forests. Similarly, Brian Geils talked about a common garden study at Fort Valley Experimental Forest with Northern Arizona University and others. Stan Kitchen told of another common garden study at Desert Experimental Range looking at the adaptive range of blackbrush (Coleogyne ramosissima), a desert shrub.

Outside of common garden studies, participants described several other types of adaptation-related research. Bob Denner mentioned experimental plantings in logged areas on Priest River Experimental Forest. Kelly Elder discussed experiments looking at hydrological responses to forest manipulations and climate change on Fraser Experimental Forest. Michael Ryan told of experiments to test the ecological impacts of one potential adaptation, fuel treatments, on Manitou Experimental Forest. Meanwhile, Terrie Jain discussed her work trying to create resilience in forests at Deception Creek Experimental Forest through experimental silvicultural treatments, while taking into account the social context in terms of what silvicultural treatments are socially acceptable.

While there was relatively little existing research related to climate change adaptation and only a handful of proposals, most participants think that EFRs could play a very important role in studying climate change adaptation due to their representativeness and ability to host manipulative research.

Role of EFRs in Future Climate Change Research

Nearly all participants argued that EFRs could and should play a major role in climate change research. They suggested that EFRs are uniquely positioned to contribute to studies of both climate change impacts and adaptation actions. In making this case, they drew on many of the strengths of EFRs described earlier in this report, including (1) the long-term nature of research on EFRs, including the level of research integrity and control provided by the designation; (2) the representativeness of the EFRs and the large scale of many of the sites; and (3) the ability to do manipulative, experimental research. According to one participant, "to have these areas so they can check on climate change is very, very important."

Dan Neary, in discussing research on climate change impacts, explained that EFRs are the "place to do it because you've got long-term data. And that's the only way you're going to really be able to tell." Bob Musselman also focused on the long-term data, saying:

> We have the role of having the extremely rich database of long-term data. We can go back and look at that over different types of ecosystems and landscapes to see how things have changed and how they'll change in the future. I think that's one of the most valuable goals is to network our datasets long term... My theory is the longer you monitor data, the more valuable it becomes. So 20 years is good. Thirty years is better. Forty years is better. Ten years is okay.

The value of long-term data was described again and again. Wyman Schmidt argued that the long-term nature of climate change requires baseline data, which can be most easily supplied by the EFRs:

> The main thing, like we've mentioned before, because climate change is so long-term, you know we're talking about the movement of glaciers, you have to establish some baseline information somewhere along the line, and Experimental Forests are the logical place to do it in our area.

For many participants, the long-term data in combination with the research integrity provided by the level of control over the landscape of EFRs made them an ideal site for climate change research. Durant McArthur argued that this control provides for consistency:

> To understand changes in climate change impacts, you need longterm studies in places that can be monitored consistently. And these kind of assets [EFRs], which the agency has more control over, are very important in those kinds of data collection.

Ward McCaughey also emphasized the control on EFRs as an important asset for researching climate change impacts:

> I think they're all important. All the Experimental Forests are important because there is very tight control over that area. And that area is, in almost all cases, probably stuck in the middle of National Forest land that surrounds it. And on National Forest land they're constantly cutting, changing, converting. And these islands of controlled environments I think are very important long-term climate change for monitoring effects. When you do make a change, say within the Experimental Forest close to that site, you can at least attenuate for it or at least know when it occurs, exactly when it occurs and monitor for that direct change.

Many participants described this long-term control as research integrity and argued that it represented a real asset for topics that require a long-term approach, such as climate change.

In addition, participants argued that EFRs were uniquely positioned to contribute to research on climate change impacts and adaptation strategies because they represented a diversity of ecosystem types. Participants used several terms to describe this strength, including the "gradient" or "network" of EFRs. Bob Musselman discussed the strength of the EFR network for looking at how species and ecosystems adapt to climate change:

> We have such a wide variety of ecosystems and habitats where we have Experimental Forests where we're doing this monitoring and that's pretty valuable as a network rather than just as an answer for that one site. So we're in a really good position to show how systems will adapt to climate change because we're looking at it from all our systems.

Terrie Jain argued that the gradient over which EFRs occur makes them important for looking at adaptation. Michael Ryan described a slightly different aspect of representativeness. He explained that the EFRs comprise the majority of the geographic range of several species, making them good areas to do manipulative experiments to study the resilience of those species under different conditions.

In our Station here, we have Experimental Forests that cover sort of the gradient in a couple of species, like say most notably ponderosa pine. So we go all the way from Arizona up to Idaho I think. And I think they'd be really good places to both do sort of simple common experiments about global change effects like snow manipulation or precip. manipulation and also to look at those in conjunction with some adaptation options, whether it's planting species that you might expect to be there in the future or doing treatments to lower density to make the surviving trees a little bit more resilient. We're set up to do stuff like that. And we could do stuff like that.

As noted previously, participants noted that the number of studies looking at adaptation strategies was limited. According to Wayne Shepperd, this lack of adaptation-related research may be due to a bias against active management:

There was a lot of climate change research underway modeling as well as some observational stuff, some experimental stuff. But a lot of that wasn't translated to management actions. I think there's a bit of a bias there. A lot of ecologists really aren't into management. I find it ironic that recently we're beginning to hear in print that they're advocating we need to actively manage. And I see this also in some conservation and environmental groups accepting that some management might be appropriate. I think this is good. But there wasn't a lot of research looking at how can we manage for climate change. But as I just said, I think we need to start doing that. And we are. But, I think, the future of research will be looking at the response to disturbances and how we might help that by tweaking the system. In many cases, preserving or at least sustaining for a longer period of time existing forest vegetation might require density control. What's limiting to a lot of the future distribution of forests isn't the ability of trees to survive in an environment. It's the abilities for trees to establish in that environment. And so initially, we won't see new seedlings. But we can keep those older trees around by maybe some judicious thinning, perhaps. That's a hypothesis that needs testing in a lot of areas under a lot of circumstances. So it's not a hands-off approach. I think a lot of people recognize that is okay... It gets back to how can we look at the underlying factors involved and then formulate a hypothesis that makes sense based on what we know about the basic ecologic requirements for species. And so I think we'll move that direction. I don't think we have yet. There isn't history of doing that, if you will, but there's a lot of work that's been done that could be used getting back to that theme.

In keeping with Wayne's discussion of active management, many participants cited the ability to do manipulative research as a unique feature of EFRs, one that could contribute a lot to the study of adaptation strategies. According to Linda Joyce, "They are also great places to try and test out some of these adaptation options for which we don't really have much precedent." Another participant argued that the ability to manipulate was what separated EFRs from research natural areas (RNAs).

> The advantage to Experimental Forests over research natural areas is that you can do manipulative research on Experimental Forests and you can't in an RNA. I think they're a prime location for that kind of work. And I would be willing to bet you that if the scientists in charge all had a chance to jump at some networked common garden or something like that, they would all love to do it if it was funded.

Stan Kitchen, in thinking about the contribution of EFRs to climate change, made a similar argument about the importance of experimentation:

> They are Experimental Forests, Experimental Ranges, which means we expect to experiment. They're not necessarily meant to be just another location that is, like a research natural area, untouched. Many of them have research natural areas, including the Desert and one very close to the boundaries of the Great Basin Experimental Range. They have their purpose, they're your control, they're your contrast. But I absolutely think that we have the opportunity to do that experimentation.

Other participants emphasized the important of RNAs, suggesting the permanent plots in RNAs provide "information about succession of vegetation in old-growth forests and ranges... points to initiate research following disturbance, especially fire." Another example came from Wayne Shepperd, who argued that the manipulative capacity of EFRs makes them good demonstration sites for experiments looking at adaptation strategies:

> They provide a good demonstration area. This is the area where you can set up an experiment and set it up as a demonstration to test a theory or to answer a question. And then you can bring people to see that on, in most cases, a relatively small scale. On some cases an operational scale. But that, in turn, can lead to further operational scale tests within the National Forest System.

Scale was also important in responses regarding adaptation research because the EFRs, while capable of doing larger-scale research than many other research sites, cannot typically host landscape-level research, which some participants claimed would be necessary for adaptation. Several participants also mentioned that capacity to do adaptation research was variable across individual EFRs because of their size and their history of research. Along these lines, participants also mentioned that it would probably be necessary to expand research outside of EFRs so that results could be extrapolated reliably to broader landscapes. Participants argued that these traits were important for adaptation research because, especially in forest environments, experiments involving adaptation treatments would necessarily be long term, requiring decades to show results.

In summary, participants argued that EFRs could and should play a "huge role" in climate change research, based on the long-term data, "the stability on the landscape where we can control what happens," and the ability to do manipulative experiments.

Challenges: Funding, Commitment, and Tough Decisions

While nearly all participants felt that EFRs play an important role in looking at climate change, many voiced concerns about potential challenges or limitations. In thinking about the potential contribution of EFRs to knowledge of climate change impacts and adaptation strategies, participants once again focused upon the institutional context within which the EFRs operate. Thus, the comments in this section largely echo the views described earlier on institutional context and change.

One participant suggested that maintaining "Experimental Forests [to] provide that outdoor laboratory where you can do experiments on a variety of scales with the good control" would be required in order to realize the potential of EFRs to contribute to climate change. Another participant, echoing sentiments described by a number of participants, felt that there was not enough commitment to the EFRs to allow them to play a big role in looking at climate change:

> But in a general sense, I think [EFRs are] a tremendous asset. There are data that could be used, but it will require some level of commitment that I don't, within this Station... I don't see the commitment there.

Experimental Forests and Climate Change   59

Some participants suggested that RMRS was increasingly recognizing the EFRs, while others felt that there was a lack of commitment that would hamper the ability of the scientists to do climate change-related research on EFRs.

Many EFR employees associated lack of commitment with lack of funding. In fact, the most common recommendation that participants gave for ensuring that the EFRs would be able to look at climate change impacts and adaptation strategies was for RMRS to provide enough funding for the EFRs to continue operating without downsizing. Participants emphasized that long-term research is not cheap and that financial commitments need to be drawn up for time periods longer than fiscal years because the EFRs need long-term stability to do long-term research. One participant explained that funding is the deciding factor in the ability of the EFRs to look at climate change impacts and adaptation: "It always boils down to, 'what can we afford to do?'" Another participant, in thinking about potential climate change research, explained that the funding situation of EFRs is dire:

> The EFRs are on life support right now as far as funding. We don't have enough money to do what we have to do, our staff, our facilities... The Experimental Forests are in danger of going away because we don't have the baseline support to keep them going. We're struggling. We're on a shoestring budget and barely getting by. That's my thought... The basic infrastructure for Experimental Forests is expensive. And it's just a matter of what the commitment of the Research Station and the national system is to maintain that long-term database that I think is so valuable. I don't know if they value that enough to continue to support it or increase the support where we can survive or not. We're barely getting along where we are now.

One participant suggested the some EFRs had such a low level of support that he could not envision them playing much of a role at all, saying "it's hard visualize [this EFR] having much of an input with the lack of work that's being done there currently. And I don't mean to disparage the people that are working. It's just that they aren't funded to do it." The following participant made the case that funding will be required to homogenize data collection on the EFRs in order to make them capable of looking at climate change impacts.

> The interesting thing about the Experimental Forests is they all were these independent entities. And now we're moving towards a network. One of the drawbacks of being independent entities is that

we weren't necessarily collecting all of the same data, or even for meteorology we weren't collecting all of the same meteorological variables on all the Experimental Forests. And we're still not. Unfortunately, getting a homogeneity to the data collection requires more funding, and we're in a less funding environment.

Another participant echoed this concern, explaining that it may be difficult to compare across the long-term datasets of different EFRs because data collection procedures were not the same. The following participant suggested that some climate change research opportunities had been missed due to the lack of funding:

> I think there were some opportunities that we couldn't capitalize on largely because of funding, in terms of instrumenting a lot of these areas, that given what we think we know now relative to climate change or anything else, just year-to-year fluctuations that we really, we've missed the boat. We've missed some opportunities.

The implication here is that data collection opportunities, once lost, may not be regained. Other participants argued that funding for facilities was necessary in order to draw scientists to the EFRs. In many cases, the need for additional funding was discussed alongside the need for more recognition for the EFRs.

EFR employees also acknowledged the need for the Station to make tough decisions in the face of budget shortfalls. They discussed the challenges of prioritizing and acknowledged the many trade-offs inherent in budget decisions. They also argued for balance in terms of funding priorities. One participant called it a "big strategic problem" saying:

> There's a challenge and there's an opportunity... I think the challenge remains of figuring out how to integrate and value all these different kinds of research both in the short term and the long term and especially in these times of economic scarcity that people have a lot, have a tendency, it's just natural human nature, to become more protective of their piece and maybe not be reaching out to others.

He acknowledged that Station employees might advocate for their own programs in a time of scarcity, but that decision-makers are challenged to find ways to value different kinds of research. Another participant also acknowledged the challenges of this level of responsibility:

> They have a large responsibility, all kinds of science to fund. But this should be a part of the mix. I believe [EFRs] should be maintained, not necessarily at the expense of other parts of the program, but the Station's programs shouldn't squeeze these valuable assets out. So I guess what I'm saying is let's have reasonable balance as we go forward.

Several participants argued for balance as they thought through the role of EFRs in climate change research and the funding required to realize the potential contributions of EFRs. One participant suggested that RMRS develop priorities with each program, identify which EFRs would be most critical to those priorities, and allocate funding accordingly.

A few participants discussed the possibility of closing some EFRs. Several participants mentioned feeling that the RMRS station leadership had been trying to avoid any such drastic measures and, as a result, all of the EFRs were suffering and becoming less effective. These participants contended that the Station would be better off if it focused funding on a smaller number of EFRs or projects, thereby ensuring that at least some of the EFRs would retain enough funding to continue research. One participant acknowledged that some EFRs may need to be closed in order for the rest to effectively do climate change research:

> I realize in tight times that this might require that some may have to be closed. And this is not unprecedented. We had Experimental Forests that no longer exist.

Another participant described this potentially difficult decision, saying:

> We're going to have to make certain tough decisions. I'm totally against closing things, long-term data collection efforts, whether it's a USGS gauge, river gauge, or an Experimental Forest, because once you close them, it's a done deal. And the investment in them goes out the door. They're worthless overnight. But at the same time, if the whole system is dying and becoming worthless as a whole, maybe it's time to make some tough decisions and close some...

Other participants strongly opposed closing any EFRs, arguing for the value of each site based on the long-term datasets, the representativeness of the EFRs, and their research integrity.

Summary of the Role of EFRs in Climate Change Research

EFR employees described numerous ways that EFRs are and could potentially contribute to improved understanding of climate change impacts and adaptation actions.

- Long-term studies can be used to answer climate change questions

Because EFRs are the sites of numerous long-term data collections efforts, they are positioned to answer many climate change questions through existing data collection. While many existing studies were not originally designed to address climate change, depending on the nature of the measurements, researchers may be able to answer important climate change questions with existing/ongoing datasets. For example, atmospheric data can be combined with information on particular tree species to understand how forests are responding to changing conditions. New long-term studies can also be initiated to answer emerging climate change questions.

- Research integrity and level of control makes EFRs important sites for climate change research

Because researchers have long-term control over the management and science, studies on EFRs have an uncommon level of research integrity. Since research on climate change impacts and adaptation actions likely requires a long-term approach, this level of research integrity is particularly important because it assures scientists that studies can continue into the future with a minimum of interference.

- EFRs are uniquely positioned for experimental/manipulative studies

Experimental studies that manipulate ecosystems may be necessary for understanding both the ways that ecosystems and species are adapting to climate change and the efficacy of different adaptation strategies. From common garden studies to larger-scale treatments to increase resilience, EFRs offer sites that can accommodate these manipulations and monitor outcomes over the long term.

- EFRs represent a range of ecosystems allowing for climate change research that applies to most western landscapes

Because of the diversity of EFR ecosystems, they can be viewed as a sort of network or gradient, representing the range of ecosystems that are being impacted by and are responding to climate change.

- Realizing the potential of EFRs to contribute to climate change research requires commitment, funding, and tough decisions

Realizing the potential of EFRs to contribute to improved understanding of climate change impacts and adaptation actions requires long-term commitment, adequate funding to maintain sites and long-term data collection and management, and difficult decisions regarding how to allocate funds in a budget shortfall.

IMPLICATIONS AND RECOMMENDATIONS

The ecological changes identified by EFR employees are not unique to EFRs; they are the shifts in forests and other ecosystems that are being observed and documented in similar landscapes throughout the Interior West. What is unique is the ability of the EFRs to contribute to our understanding of these changes and the efficacy of possible management responses. Here, we summarize the key insights and recommendations made by 21 EFR employees in their in-depth interviews on climate change and EFRs and provide some additional interpretations and recommendations.

- EFRs build on a history of relevance, science leadership, and application value. They are uniquely positioned within the Forest Service and within the world to ask and answer important questions in natural resource management. The key challenge is how to maintain and build the potential of EFRs in a constrained budget environment.
- Informal, field observations play an important role in research. These observations generate ideas for projects and specific research questions and improve the interpretation of results. To the extent that researchers are able to spend significant time in the field and develop long-term relationships with particular EFRs, field observations will continue to improve the quality of EFR research.
- Long-term data collection is one of the hallmarks of the EFRs and is particularly important in the context of climate change. Understanding climate change impacts and the efficacy of adaptation actions requires

a long-term approach and long-term investment in data collection efforts. Importantly, both existing and new long-term studies can contribute to knowledge of climate change if adequate support can be secured. In some cases, investment in long-term data collection may require valuing datasets whose future uses are certain but not yet specified.

- EFRs offer a potentially unparalleled level of research integrity as well as the ability to manipulate/conduct treatments at multiple scales. Climate change research, particularly research on adaptation actions that are largely untested, can benefit from experimental treatments in a relatively controlled environment in the long term. It may be important to identify mechanisms to facilitate experimental treatments, for example through additional support for NEPA review or expedited review processes.

- While the EFR employees in this study did not focus on partnerships as a mechanism to build capacity, they described numerous collaborative studies with university partners and the potential for EFRs to become part of larger networks. Partnerships may provide a means to leverage additional resources for EFRs. EFRs can build on existing research priorities throughout the RMRS, they can connect to existing monitoring networks across North America, and they can partner with universities and non-governmental organizations that have access to other pools of federal dollars. In particular, promoting EFRs to potential university collaborators as a site where experimental treatments can be conducted in relatively controlled environments over the long-term might attract new partners with different types of resources.

- Universities and state agencies could also become long-term partners of specific EFRs, where desirable. Memoranda of Understanding could facilitate collaborations that allow these non-Forest Service institutions to invest in EFRs for the long-term. Andrews Experimental Forest provides an interesting such model of Forest Service-university collaboration.

- EFRs exist within a broader institutional context–the RMRS, Forest Service research, and North American science–all of which are experiencing changes in terms of levels of funding and the way funding is allocated (e.g., competitive proposals for short-term research versus long-term investments in research programs). Realizing the potential of EFRs to contribute to climate change may require swimming against the current to fund long-term studies and invest in specific sites.

ACKNOWLEDGMENTS

We thank (1) the research participants for their thoughtful interview responses and careful reviews of this report, (2) the USDA Forest Service Rocky Mountain Research Station for its support of and interest in this project, and (3) the external reviewer who provided additional insights and valuable feedback on the role of EFRs.

REFERENCES

Adams, Mary Beth, Linda Loughry, and Linda Plaugher. 2008. Experimental Forests and Ranges of the USDA-Forest Service. General Technical Report NE-321 Revised. Newtown Square, PA: USDA-Forest Service, Northeastern Research Station.

Bart, David. 2006. Integrating Local Ecological Knowledge and Manipulative Experiments to Find the Causes of Environmental Change. Frontiers in Ecology and the Environment 4(10):541-546.

Breshears, David, Neil Cobb, Paul Rich, Kevin Price, Craig Allen, Randy Balice, and others. 2005. Regional Vegetation Die-Off in Response to Global-Change-Type Drought. PNAS 102(42):15144-15148.

Cole, David, Constance Millar, and Nathan Stephenson. 2010. Pp. 179-196 in: Beyond Naturalness: Rethinking Park and Wilderness Stewardship in an Era of Rapid Change. Edited by David Cole and Laurie Yung. Washington, DC: Island Press.

Cook, Carly, Marc Hockings, and R.W. Carter. 2009. Conservation in the Dark? The Information Used to Support Management Decisions. Frontiers in Ecology and the Environment 8:181-186.

Fazey, Ioan, John Fazey, Janet Salisbury, David Lindenmayer, and Steve Dovers. 2006. The Nature and Role of Experiential Knowledge for Environmental Conservation. Environmental Conservation 33(1):1-10.

Frelich, Lee and Peter Reich. 2009. Wilderness Conservation in an Era of Global Warming and Invasive Species: A Case Study from Minnesota's

Boundary Waters Canoe Area Wilderness. Natural Areas Journal 29:385-393.

Gibson, Ken, Kjerstin Skov, Sandy Kegley, Carl Jorgensen, Sheri Smith, and Jeff Witcosky. 2008. Mountain Pine Beetle Impacts in High-Elevation Five-Needle Pines: Current Trends and Challenges. Publication number R1-08-020. Missoula, MT: USDA Forest Service, Forest Health and Protection.

Heller, Nicole and Erika Zavaleta. 2009. Biodiversity Management in the Face of Climate Change: A Review of 22 Years of Recommendations. Biological Conservation 142:14-32.

Millar, Constance, Nathan Stephenson, and Scott Stephens. 2007. Climate Change and Forests of the Future: Managing in the Face of Uncertainty. Ecological Applications 17(8):2145-2151.

Mowrer, Todd. 2011. RMRS Experimental Forest and Range Importance and Background. Unpublished draft charter for USDA Forest Service Rocky Mountain Research Station Experimental Forests and Ranges.

Parmesan, Camille. 2006. Ecological and Evolutionary Responses to Recent Climate Change. Annual Review of Ecology, Evolution, and Systems 37:637-69.

Parmesan, Camille and Gary Yohe. 2003. A Globally Coherent Fingerprint of Climate Change Impacts across Natural Systems. Nature 421:37-42.

Robertson, Hugh and Tara McGee. 2003. Applying Local Knowledge: The Contribution of Oral History to Wetland Rehabilitation at Kanyapella Basin, Australia. Journal of Environmental Management 69:275-287.

Samberg, Leah. 2011a. Climate Change and Species Conservation. Science Synthesis for Wilderness.net. Produced by the Aldo Leopold Wilderness Research Institute and the University of Montana.

Samberg, Leah. 2011b. Climate Change and Ecosystem Management. Science Synthesis for Wilderness.net. Produced by the Aldo Leopold Wilderness Research Institute and the University of Montana.

Samberg, Leah. 2011c. Climate Change and Invasive Species. Science Synthesis for Wilderness.net. Produced by the Aldo Leopold Wilderness Research Institute and the University of Montana.

Samberg, Leah. 2011d. Climate Change, Fire, and Ecosystems in the US. Science Synthesis Wilderness.net. Produced by the Aldo Leopold Wilderness Research Institute and the University of Montana.

Stephenson, Nathan, Constance Millar, and David Cole. 2010. Shifting Environmental Foundations: The Unprecedented and Unpredictable Future. Pp. 50-66 in: Beyond Naturalness: Rethinking Park and

Wilderness Stewardship in an Era of Rapid Change. Edited by David Cole and Laurie Yung. Washington, DC: Island Press.

Wells, Gail, Deborah Hayes, Katrina Krause, Ann Bartuska, Susan LeVan-Green, Jim Anderson, and others. 2009. Experimental Forests and Ranges: 100 Years of Research Success Stories. General Technical Report FPL-GTR-182. Madison, WI: USDA-Forest Service, Forest Products.

Young, Jeremy. 2008. Roots of Research: Raphael Zon and the Origins of Forest Experiment Stations. Pp. 363-370 in: Fort Valley Experimental Forest: A Century of Research 1908- 2008, RMRS-P-53CD edited by Susan Olberding and Margaret Moore. Fort Collins, CO: USDA-Forest Service, Rocky Mountain Research Station.

APPENDIX A: EXPERIMENTAL FORESTS AND RANGES IN THE ROCKY MOUNTAIN REGION

Experimental Forests and Ranges (EFRs) administered by the Rocky Mountain Research Station (RMRS) include a wide spectrum of landscapes, productivity, and elevations. This section gives a basic introduction to the ecology, history, and research focus of each. There are three EFRs in Idaho: Priest River, Boise Basin, and Deception Creek EFs; two in Montana: Coram and Tenderfoot Creek EFs; one in South Dakota: Black Hills EF; one in Wyoming: Glacier Lakes Ecosystem Experiments Site; two in Colorado: Manitou and Fraser EFs; three in Arizona: Sierra Ancha, Long Valley, and Fort Valley EFs; and two in Utah: Great Basin and Desert ERs. We present them in alphabetical order.

Black Hills Experimental Forest, located in Black Hills National Forest in western South Dakota, was established in 1961. Black Hills EF was designated for research in multi-use management of ponderosa pine forests and continues to offer research opportunities for studies in ponderosa pine with a wide range of age classes and stand structures owing to silvicultural treatments performed in the 1980s. Black Hills has an Ameriflux tower that is operated by the South Dakota School of Mines.

Boise Basin Experimental Forest is located adjacent to the city of Idaho City, Idaho, and was established in 1933. Boise Basin's original research focus was ponderosa pine management along with the ecological effects of silvicultural treatments. It has long-term databases on fire return intervals, soils, habitat types, and vegetation trends. Current research opportunities include the wildland-urban interface, recreation, and ponderosa pine restoration. Boise Basin has hosted collaborations between RMRS scientists and the National Interagency Fire Center, Idaho University, Forest Service Region 4, Boise National Forest, and the Boise Cascade Corporation.

Coram Experimental Forest, in northwest Montana, was also established in 1933. Coram, along with Glacier and Waterton Lakes Parks, is part of the Crown of the Continent Biosphere Reserve. Research there has been focused on western larch regeneration and management and includes long-term databases on natural regeneration following different silvicultural treatments and site preparations, small mammal interactions in larch forests, conifer seed dispersal, and direct seeding of conifers. Coram offers a wide variety of age classes and stand types owing to its history of experimental silvicultural treatments and site preparations. RMRS researchers at Coram have collaborated with researchers from a wide range of national and international universities and land agencies.

Deception Creek Experimental Forest is located 32 kilometers east of Coeur d'Alene, Idaho, and was established in 1933 for silvicultural research in western white pine management. Located in an area with abundant precipitation, Deception Creek has probably the highest productivity of any EFR in RMRS territory. It has long-term databases on forest growth, weather, and western white pine genetics. Because of a variety of stand ages, Deception Creek offers a wide range of conditions for research. Collaborators include area universities, Forest Service Region 2, and the Idaho Panhandle National Forest.

Desert Experimental Range, located in western Utah, was established in 1933 to serve as a center for rangeland research. It was designated as a Biosphere Reserve in 1976 and is the only such Reserve in the western hemisphere in a cold-desert landscape. Desert ER has long-term databases on daily precipitation and temperature, vegetation maps, and grazing exclosures. Research has been focused on disturbance and successional processes in cold-desert environments, desertification, winter sheep management, rodent ecology, pronghorn antelope biology and management, cryptobiotic soil-crust ecology, and avian and mammalian population dynamics.

Fort Valley Experimental Forest received EFR designation in 1931, though research began there in 1908, making it the earliest experimental site in the National Forest System. Located 14 km north of Flagstaff, Arizona, Fort Valley sits at 2176 m elevation in a ponderosa pine forest. Long-term databases include weather, ponderosa pine regeneration, range research, mistletoe research, and racial variation in ponderosa pine. Current research

Experimental Forests and Climate Change   69

directions include forest pathology, forest restoration, forest management in the wildland-urban interface, and fire effects. Collaborators at Fort Valley include Coconino National Forest, Northern Arizona University, the Ecological Restoration Institute, and a local school district.

Fraser Experimental Forest is located 112 km northwest of Denver, Colorado. The environment ranges from alpine to subalpine, with Engelmann spruce and lodgepole pine forests and alpine tundra. It was established in 1937 for research into the effects of forest management on water yields and timber production.

Fraser has long-term databases on streamflow, snowdepth and water content, meteorology, seed production, and tree growth and mortality. Collaboration at Fraser includes the USDI Geological Survey, NASA, National Oceanic and Atmospheric Administration National Weather Service, National Center for Atmospheric Research, and many universities.

Glacier Lakes Ecosystem Experiments Site (GLEES) in Wyoming was established around 1990 to research the effects of atmospheric deposition and climatic change on alpine ecosystems. GLEES is considered a part of the EFR network, but it does not yet have an official EFR establishment document. This is in the works, according to one of the study participants. GLEES is part of several national monitoring networks, including the Ameriflux system and the National Atmospheric Deposition Program. Long-term databases include climate, atmospheric chemistry, water chemistry, snow, and vegetation.

Great Basin Experimental Range was established in 1912 in response to catastrophic flooding. It is located east of Ephraim, Utah, on the west side of the Wasatch Plateau in the Manti-La Sal National Forest. One of the locations for seminal research on range management, Great Basin's research emphasis has also included watershed ecology and management and silviculture. Long-term databases include climate, streamflow, restoration plantings, and the effects of grazing treatments and exclosures. Current research directions include plant adaptation and successional processes, nutrient cycling, restoration ecology, and game habitat improvement.

Long Valley Experimental Forest was established in 1936 to represent a contrasting soil type to that of Fort Valley Experimental Forest. Located 96 km southeast of Flagstaff, Arizona, Long Valley is primarily composed of ponderosa pine stands and has hosted research with a focus on burning, thinning, and regeneration.

Manitou Experimental Forest is located in Colorado's Front Range near Colorado Springs. It was designated in 1936 with a primary research focus on the restoration of pine-grassland ecosystem.

In 1980, the research emphasis shifted to ponderosa pine management and wildlife. It has also served as an important location for research into the wildland-urban interface because of its proximity to urban centers and the numerous private inholdings it contains. Long-term databases include weather, deposition, streamflow, and information on the flammulated owl. Manitou has hosted collaborative research with several Colorado universities and the USDI Geological Survey.

Priest River Experimental Forest, established in 1911, was one of the earliest sites to receive EFR designation. It has long-term weather, forest growth, stream-flow, and snowfall databases and has been a site for important work in silviculture, fire, and forest genetics. Its facilities and laboratories have allowed it to host scientists and research projects, leading to collaborations with area universities, state land management agencies, and Forest Service districts.

Sierra Ancha Experimental Forest was established in 1932 and is located along the eastern slope of the Sierra Ancha mountain range approximately 175 km northeast of Phoenix, Arizona. The primary plant associations are chaparral shrubs, but there are also mixed conifer stands, oak woodlands, and grasslands, among others. Sierra Ancha's original research focus was watershed management, and it has hosted collaborations with the Tonto National Forest, Arizona universities, and the Salt River Water Users Association.

Tenderfoot Creek Experimental Forest, located in the Little Belt Mountains east of the Continental Divide in Montana, was established in 1961 to study the management of lodgepole pine forests. Long-term databases include information on timber inventories, streamflow and water quality, and fuels analysis, among others.

Beginning in 1991, Tenderfoot Creek has been host to a long-term research project on different options for managing lodgepole pine forests, meaning that there are many opportunities for current and ongoing studies in this area.

Scientists at Tenderfoot Creek have collaborated with several National Forests, area universities, and land management agencies.

APPENDIX B: ROCKY MOUNTAIN RESEARCH STATIOIN EFR DATA NETWORK

	Met	Hydrol	ICP2	NADP	Ameri-flux	HydroDB	ChemDB	SnoTel	CASTNET
Black Hills					X*				
Boise Basin									
Coram	X	X							
Deception Creek									
Desert ER	X								
Fort Valley	X							X	
Fraser	X	X				X			
GLEES	X	X	X	X	X			X	X
Great Basin ER	X								
Long Valley									
Manitou	X				X*				
Priest River	X	X							
Sierra Ancha	X	X	X						
Tenderfoot	X	X	X					X	

*Maintained by cooperators.

Key:

Met = EFR collects long-term meteorological data.

Hydrol = EFR collects long-term hydrological data.

NADP= National Acid Deposition Program, http://nadp.sws.uiuc.edu/

ICP2 = International Co-operative Program on Assessment and Monitoring of Air Pollution Effects on Forests, http://icpforests.net/

Ameriflux = The AmeriFlux network provides continuous observations of ecosystem level exchanges of CO_2, water, energy and momentum spanning diurnal, synoptic, seasonal, and interannual time scales and is currently composed of sites from North America, Central America, and South America, http://public.ornl.gov/ameriflux/

HydroDB/ClimDB = centralized server to provide open access to long-term meteorological and streamflow records from a collection of research sites, http://www.fsl.orst.edu/climhy/

SnoTel = automated system to collect snowpack and related climatic data in the Western United States called SNOTEL (for SNOwpack TELemetry), http://www.wcc.nrcs.usda.gov/

CASTNET = The Clean Air Status and Trends Network (CASTNET) is a national air quality monitoring network designed to provide data to assess trends in air quality, atmospheric deposition, and ecological effects due to changes in air pollutant emissions, http://epa.gov/castnet/javaweb/index.html.

APPENDIX C: INTERVIEW GUIDE

What is/was your role on the _____ Experimental Forest/Range? (How long have been working/did you work with _____ EF/R)? What does/did your job entail relative to the _____ EF/R?)

What kind of time have you spent/did you spend in the field in? _____

How would you describe the _____ Experimental Forest/Range to a member of Congress? What unique or special contribution does this Experimental Forest/Range make?

When you first started here, what were some of your early impressions of this Forest/Range? What changes have you seen on the Forest/Range since you started working here?

What sort of institutional changes have you seen during this time? I'm thinking about changes in the program of work, staffing, funding, and so on. How have those changes influenced research on the _____Experimental Forest/Range?

What role do/did your regular, day-to-day experiences in the field at_____ Experimental Forest/Range over the _____ years you've/you worked with this forest/range play in your understanding of change ?_____

What role do/did these day-to-day experiences in the field at _____ play in the development and interpretation of experiments/research?

Can you tell me about the relationship between the _____ Experimental Forest/Range and nearby communities or other forest users?

Going back to some of the ecological changes you described earlier, what do you think is causing those changes? Are there changes that you think are happening because of climate change?

How have these changes affected the program of work or focus of the _____ Experimental Forest/Range?

Can you tell me about the climate change-related research on this Forest? - if answer addresses only climate monitoring-type research probe about impacts-related research

Are you conducting or planning any experiments/research to test proposed climate change adaptation actions? I'm thinking of adaptation as actions that land managers take to assist ecosystems and species in adapting to new conditions.

What role do you think Forest Service Experimental Forests and Ranges in general could and should play in advancing our understanding of climate change impacts?

What role do you think they should play in testing adaptation actions and advancing climate change adaptation?

How do you think the Forest Service, or more specifically the Rocky Mountain Research Station, should support and utilize the Experimental Forests and Ranges to ensure that they are able to fill those roles?

Is there anything else you'd like to add?

APPENDIX D: INTERVIEW PARTICIPANTS

Name	Experimental Forest/Range(s)	Occupation	Years with EFRs
Brian Geils	Fort Valley, Manitou, GLEES, Fraser, Priest River, Deception Creek	Scientist-in-charge (former) and researcher	30
Dan Neary	Fort Valley, Long Valley, Sierra Ancha	Scientist-in-charge and researcher	18
Carl Edminster	Manitou, Fort Valley, Long Valley, Fraser, Black Hills, Sierra Ancha	Lead scientist and researcher (retired)	25
Michael Ryan	Manitou, Fraser	Scientist-in-charge and researcher	22
Wayne Shepperd	Fraser, Manitou, Black Hills	Manager and researcher (retired)	36
Chuck Troendle	Fraser, Manitou	Administrator and researcher (retired)	22
Kelly Elder	Fraser	Scientist-in-charge and researcher	10
Linda Joyce	Manitou, GLEES, Black Hills	Administrator	14
Bob Musselman	GLEES	Scientist-in-charge and researcher	22
Russ Graham	Deception Creek, Priest River, Boise Basin, Black Hills	Scientist-in-charge and researcher	35

Appendix D. (Continued)

Name	Experimental Forest/Range(s)	Occupation	Years with EFRs
Terrie Jain	Deception Creek, Boise Basin, Priest River, Black Hills	Scientist-in-charge and researcher	10
Bob Denner	Priest River, Boise Basin, Deception Creek, Black Hills	Supervisory forester	20
Ward McCaughey	Coram, Tenderfoot Creek	Manager and researcher (retired)	20
Ray Shearer	Coram	Manager and researcher (retired)	46
Durant McArthur	Great Basin, Desert	Administrator, scientist-in- charge, and researcher (retired)	37
Gary Jorgensen	Great Basin, Desert	Technician (retired)	39
Stan Kitchen	Desert, Great Basin	Site manager, scientist-in- charge, and researcher	19
Wyman Schmidt	Coram, Tenderfoot Creek	Manager and researcher (retired)	35

* An additional three participants chose to remain anonymous.

In: Role of Experimental Forests and Ranges ... ISBN: 978-1-63463-729-9
Editor: Kayla Pierce © 2015 Nova Science Publishers, Inc.

Chapter 2

CLIMATE CHANGE SCIENCE: KEY POINTS[*]

Jane A. Leggett

SUMMARY

Though climate change science often is portrayed as controversial, broad scientific agreement exists on many points:

- The Earth's climate is warming and changing.
- Human-related emissions of greenhouse gases (GHG) and other pollutants have contributed to warming observed since the 1970s and, if continued, would tend to drive further warming, sea level rise, and other climate shifts.
- Volcanoes, the Earth's relationship to the Sun, solar cycles, and land cover change may be more influential on climate shifts than rising GHG concentrations on other time and geographic scales. Human-induced changes are super-imposed on and interact with natural climate variability.
- The largest uncertainties in climate projections surround feedbacks in the Earth system that augment or dampen the initial influence, or affect the pattern of changes. Feedback mechanisms are apparent in clouds, vegetation, oceans, and potential emissions from soils.

---

[*] This is an edited, reformatted and augmented version of a Congressional Research Service publication, CRS Report for Congress R43229, prepared for Members and Committees of Congress, from www.crs.gov, dated September 10, 2013.

- There is a wide range of projections of future, human-induced climate change, all pointing toward warming and associated sea level rise, with wider uncertainties regarding the nature of precipitation, storms, and other important aspects of climate.
- Human societies and ecosystems are sensitive to climate. Some past climate changes benefited civilizations; others contributed to the demise of some societies. Small future changes may bring benefits for some and adverse effects to others. Large climate changes would be increasingly adverse for a widening scope of populations and ecosystems.

As is common and constructive in science, scientists debate finer points. For example, a large majority but not all scientists find compelling evidence that rising GHG have contributed the most influence on global warming since the 1970s, with solar radiation a smaller influence on that time scale. Most climate modelers project important impacts of unabated GHG emissions, with low likelihoods of catastrophic impacts over this century. Human influences on climate change would continue for centuries after atmospheric concentrations of GHG are stabilized, as the accumulated gases continue to exert effects and as the Earth's systems seek to equilibrate.

The U.S. government and others have invested billions of dollars in research to improve understanding of the Earth's climate system, resulting in major improvements in understanding while major uncertainties remain. However, it is fundamental to the scientific method that science does not provide absolute proofs; all scientific theories are to some degree provisional and may be rejected or modified based on new evidence. Private and public decisions to act or not to act, to reduce the human contribution to climate change or to prepare for future changes, will be made in the context of accumulating evidence (or lack of evidence), accumulating GHG concentrations, ongoing debate about risks, and other considerations (e.g., economics and distributional effects).

BROAD SCIENTIFIC AGREEMENT ON MANY ASPECTS OF CLIMATE CHANGE[1]

Despite portrayals in popular media about controversies in climate change science, almost all climate scientists agree on certain important points:

- The Earth's climate has warmed significantly and changed in other ways over the past century (*Figure 1*). The warming has been widespread but not uniform globally, with most warming over continents at high latitudes, and slight cooling in a few regions, including the southeastern United States, the Amazon, and the North Atlantic south of Greenland.[2]

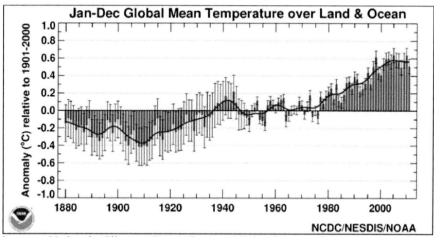

Source: National Climate Data Center, National Oceanic and Atmospheric Administration, U.S. Department of Commerce. Figure extracted March 28, 2013. Very similar findings are reported by several other, independent research groups. See, for example, Rohde, Robert, Richard Muller, Robert Jacobsen, Elizabeth Muller, Saul Perlmutter, Arthur Rosenfeld, Jonathan Wurtele, Donald Groom, and Charlotte Wickham. "A New Estimate of the Average Earth Surface Land Temperature Spanning 1753 to 2011." Geoinformatics & Geostatistics: An Overview 1, no. 1 (December 7, 2012).

Notes: Red bars represent "anomalies," or differences in mean temperature for the year compared with the 20th century average. Anomalies are a better estimate than the absolute value, as they can capture the change over time more reliably while absolute values are vulnerable to gaps in geographic coverage. The blue line shows the running average, applying a "21-point binomial filter" to the time series plotted as red bars. The "whisker" (thin black vertical) lines represent confidence or possible error levels. Confidence has improved over the past century.

Figure 1. Long-Term Temperature Observations. Compared to the 20th Century Global Mean Temperature.

- The climate has varied naturally through geologic history. Past climate changes sometimes proceeded abruptly when they passed certain "tipping points." The National Academy of Sciences concluded that the past few decades were very likely the warmest in the past 400 years, and "that temperatures at many, but not all, individual locations were higher during the past 25 years than during any period of comparable length since A.D. 900."[3] Although conclusions cannot yet be precise, research suggests that global average temperatures today are among the highest since human civilizations began to flourish roughly 4,000 to 8,000 years ago.[4]

- "Greenhouse gases" (GHG) include, among others, carbon dioxide (CO_2), water vapor, methane (CH_4), and nitrous oxide (N_2O), as well as some aerosols. They absorb energy into the atmosphere rather than letting it escape to space. The presence of GHG in the atmosphere warms the Earth to its current temperature.

- Human activities, especially fossil fuel burning, deforestation, agriculture, and some types of industry, have increased GHG concentrations in the atmosphere. CO_2, the main GHG emitted by human activities, has risen almost 40% over the past 150 years. About one-third of human-related CO_2 has been absorbed by oceans, increasing surface water acidity by 30%.[5]

- The enhanced levels of GHG in the atmosphere contributed to the observed global average warming in recent decades. Over other time and geographic scales, such factors as the Earth's orbit, solar variability, volcanoes, and land cover change have been stronger influences than human-related GHG.

- There is a wide range of projections of future human-induced climate change, with all pointing toward warming. Human-induced change will be superimposed on, and interact with, natural climate variability.

- Human societies and ecosystems are sensitive to climate. Some climate changes benefited civilizations; others contributed to some societies' demises.

- The range of possible impacts on humans and ecosystems is also very wide. In the near term, climate change (including the fertilization of vegetation by CO_2) may bring benefits for some, while adversely affecting others. Researchers expect the balance of projected climate change impacts to be increasingly adverse for a widening scope of populations and ecosystems.

As is common and constructive in science, scientists debate finer points. Some disagree with the broader consensus that GHG have been *the major* influence on global warming over the past few decades. Some suggest that, if GHG emissions continue unabated, the resulting climate change would be small and possibly beneficial overall. Most climate modelers project changes that are significant to large, with small likelihoods of changes that could be catastrophic for some human societies and ecosystems in coming decades.

DEALING WITH UNCERTAINTIES

Even the best science cannot provide absolute proof; it is fundamental to the scientific method that all theories are to some degree provisional and may be rejected or modified based on new evidence. Private and public decisions to act or not to act, to reduce the human contribution to climate change or to prepare for future changes, will be made in the context of accumulating evidence (or lack of evidence), accumulating GHG concentrations, ongoing debate about risks, and other considerations (e.g., economics and distributional effects). That said, billions of dollars have been invested in research on a wide range of climate change topics, including the possibility of attribution to alternative causes than greenhouse gases. To date, scientists have found little support for the hypothesis that GHG are not responsible for observed warming, nor have they found much evidence that other factors (including solar changes) can explain more than a small portion of global average temperature increases since the 1970s. For example, measurements of solar irradiance suggest that the solar influence on global temperatures has been decreasing overall since the 1930s, with the up-and-down pattern of the 11-year solar cycle evident in observations. A large body of research is consistent with attributing the majority of global temperature increase since the 1970s to the increase in GHG concentrations. It is this balance of evidence that leads most scientists to consider human-related GHG emissions an important global risk.

Sound Science Does Not Offer Proof

As scientists may point out, "there is no such thing as a scientific proof. Proofs exist only in mathematics and logic, not in science.... The primary criterion and standard of evaluation of scientific theory is evidence, not proof.... The currently accepted theory of a phenomenon is simply the best explanation for it among all available alternatives."6 Normal scientific methods aim at disproving a hypothesis; if evidence cannot disprove a hypothesis, it generally buttresses confidence in the hypothesis.

ISSUES FOR CONGRESS

It appears unlikely that science will provide decision-makers with significantly more scientific certainty for many years regarding the precise patterns and risks of climate change. Nonetheless, both private and public decision-makers face climate-related choices.

Broadly, response options to significant climate change include (1) defer the choices; (2) find out more; (3) inform affected populations; (4) prepare; (5) try to contain it; and (6) choose to experience the consequences. In many cases, many decision-makers are likely to face situations that require a response, such as resolving discrepancies between designated and actual flood plains or attempting to improve agricultural productivity in light of contemporary climate patterns.

Based on what is and what is not well known concerning climate change, as well as other considerations, Members of Congress may address climate-related decisions that affect

- authorizations and appropriations for federal programs, including research and technology development;
- tax and financial incentives for private decision-makers;
- regulatory authorities; or
- information or assistance to affected entities to help them adapt or rebuild after damages.

A variety of other CRS reports provide background and analysis on such options and are listed at the end of this report.[7]

Causes of Observed Climate Change: Forcings, Feedbacks, and Internal Variability

Three concepts may be useful for understanding the mechanisms and debate over the contributions to observed climate change: *forcings, feedbacks,* and *internal variability.*

Forcings

There is broad agreement among scientists that certain factors—including the composition of the atmosphere and solar variability—directly change the balance between incoming and outgoing radiation in the Earth's system and consequently *force* climate change. *Forcings* include the following:

- Atmospheric concentrations of greenhouse gas (GHG) and other trace gas and aerosol. These include water vapor,8 carbon dioxide (CO_2), methane (CH_4), nitrous oxide (N_2O), sulfur hexaflouride (SF6), chlorofluorocarbons (CFC), hydrofluorocarbons (HFC), perfluorocarbons (PFC), ozone, sulfur aerosols, black and organic carbon aerosols, and others. Human activities, especially fossil fuel burning, deforestation, agriculture, and some types of industry, have increased GHG concentrations in the atmosphere. CO_2 has risen almost 40% over the past 150 years.[9]
- Amount and patterns of solar radiation reaching the Earth due to the Earth's orbit around the Sun, and the tilt and wobble of the Earth's axis, as well as solar variability (Figure 2).
- Reflectivity of the Earth's surface due to changes in land use (e.g., urban surfaces, forest cover), changes in ice and snow cover; and vegetation cover.

Human-Related Greenhouse Gas Emissions

A majority of human-related GHG emissions are carbon dioxide, released primarily from energy production and use, deforestation and forest degradation, and cement manufacture. World-wide in 2010, carbon dioxide emissions were 74% of human-related GHG emissions. In the United States, carbon dioxide was 83% of human-related GHG emissions (*Figure 3*). Methane and nitrous oxide emissions are greater shares (16% and 8%, respectively) globally than in the United States (10% and 5%, respectively). Agriculture is a main source of these emissions and is a bigger share of the economies of many low-income countries compared with the United States. Also, many sources in the United States have acted to reduce their GHG emissions (such as in reducing leaks of methane), compared with sources in some low-income countries.

A large majority (73%) of global GHG are emitted by the 10 top emitting countries *Figure 4*). China became the leading GHG emitter in 2007 when it surpassed the United States. While China's emissions have been on the rise, the United States has emitted more cumulatively than any other country over the past 100 years.

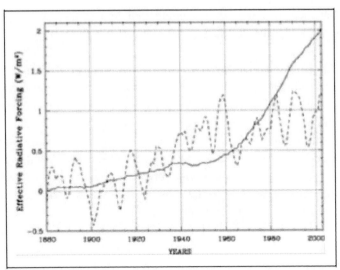

Source: Ziskin, Shlomi, and Nir J. Shaviv. "Quantifying the Role of Solar Radiative Forcing over the 20th Century." Advances in Space Research 50, no. 6 (September 15, 2012): 762–776. doi:10.1016/j.asr.2011.10.009.

Notes: According to the authors, "the optimal anthropogenic contribution (solid line) and the optimal solar contribution (dashed line) over the 20th century. The anthropogenic contribution is primarily composed of GHGs and aerosols. The solar contribution includes changes in the total solar irradiance and the indirect solar effect (ISE)." This is one of many studies, using a variety of methods, investigating the relative contributions of different climate forcings that conclude that the GHG concentrations have outweighed all other influences on global mean air surface temperature from the late 1970s to the present. For a broader, more thorough review of scientific understanding of the solar influence, see Gray, L. J., J. Beer, M. Geller, J. D. Haigh, M. Lockwood, K. Matthes, U. Cubasch, et al. "Solar Influences on Climate." Reviews of Geophysics 48, no. 4 (October 30, 2010). doi:10.1029/2009RG000282.

Figure 2. One Estimate of Human-Related versus Solar Contributions to Global Temperature Change Over the 20t[h] Century.

Feedbacks

Once a change in the Earth's climate system is underway, responses *within* the system will amplify or dampen the initiated change. Virtually all climate scientists conclude that all the feedbacks *in net* are likely to be positive (i.e., increasing climate change in the same direction caused by warming),[10] especially if temperature increases are large;[11] there remain wide differences in views.

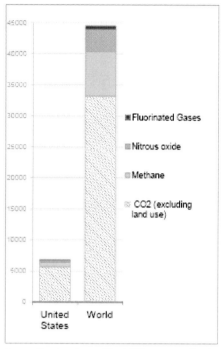

Source: CRS figure using estimates from World Resources Institute, CAIT version 2.0, extracted September 12, 2013.
Notes: These estimates cover the six GHG covered by the Kyoto Protocol (CO_2, CH_4, N_2O, SF_6, HFC, and PFC), expressed in their equivalencies to the effect of CO_2 on "radiative forcing" of the atmosphere over a 100-year period.
The World column includes U.S. emissions.

Figure 3. Shares of Human-Related GHG Emissions by Gas in 2010. Million metric tons of CO_2-equivalent.

An important consideration is that, once positive feedbacks begin, they may be essentially irreversible and, at least theoretically, lead to "runaway warming." A few of the major feedbacks are clouds, vegetation, snow and ice cover, and uptake or releases of GHG by soils and water bodies. Forests, for example, provide both negative and positive feedbacks. On the one hand, higher CO2 concentrations in the atmosphere tend to fertilize their growth (if other conditions are not limiting) and forests may grow more rapidly with greater warmth and precipitation; these factors could dampen initiated warming. On the other hand, forests thrive within certain bounds of growing conditions; if their climate conditions change beyond those bounds, they are likely to grow more slowly and eventually die back, releasing the carbon they and forest soils store and enhancing the initiated climate change.

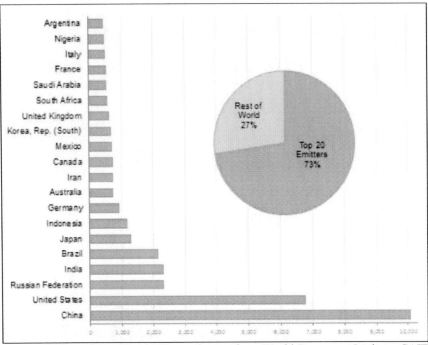

Source: CRS graphic with emission estimates from World Resources Institute, CAIT Version 2.0, extracted September 12, 2013.
Notes: These estimates cover the six GHG covered by the Kyoto Protocol (CO_2, CH_4, N_2O, SF6, HFC, and PFC), expressed in their equivalencies to the effect of CO_2 on "radiative forcing" of the atmosphere over a 100- year period.

Figure 4. Estimated Top 20 Emitting Nations of Greenhouse Gases in 2010. Million metric tons of carbon-dioxide equivalent, includes all net land use fluxes.

Internal Variability

The climate exhibits its own rhythms, or *internal variability*. The oscillation between El Niño and La Niña events is an example of internal climate variability that has important effects on economies and ecosystems in the Pacific basin (including across the United States). Another is the North Atlantic Oscillation. Internal variability may be difficult to distinguish from decadalscale climate change. Such patterns of variability also may be influenced by climate change.

PROJECTIONS OF FUTURE HUMAN-INDUCED CLIMATE CHANGE

Most climate science experts project that if GHG emissions are not reduced far below current levels, the Earth's climate would warm further, above natural variations, to levels never experienced by human civilizations. If, and as, the climate moves further from its present state, it would reconfigure the patterns and events to which current human and ecological systems are adapted, and the risk of abrupt changes would dramatically increase.

Scenarios of future GHG concentrations under current policies range from 500 ppm carbon dioxide equivalents[12] (CO2e) to over 1,000 ppm CO2e by 2100. These are projected to raise the global average temperature by at least 1.5° Celsius (2.7° Fahrenheit) above 1990 levels,[13] not taking into account natural variability. The estimates considered most likely by many scientists are for GHG-induced temperature increases around 2.5 to 3.2° C (4.5 to 5.8° F) by 2100.[14] There is a small but not trivial likelihood that the GHG-induced temperature rise may exceed 6.4°C (11.5° F) above natural variability by 2100.[15]

As context, the global average temperature at the Last Glacial Maximum has been estimated to be about 3 to 5° C (5.4 to 9° F) cooler than present,[16] and is estimated currently to be approaching the highest level experienced since the emergence of human civilizations about 8,000 years ago.[17]

Future climate change may advance relatively smoothly or sporadically, and some regions are likely to experience more fluctuations in temperature, precipitation, and frequency or intensity of extreme events than others. Almost all regions are expected to experience warming; some are projected to become warmer and wetter, while others would become warmer and drier. Sea levels could rise due to ocean warming alone on average between 7 and 23 inches by 2100. Adding to that estimate would be the effects of poorly understood but possible accelerated melting of the Greenland or Antarctic ice sheets. Recent scientific studies have projected a total global average sea level to rise in the 21st century, depending on GHG scenarios, ice dynamics, and other factors, in the range of 2 to 2.5 feet, with a few estimates ranging up to 6.5 feet.[18] Continued warming could lead to additional sea level rise over subsequent centuries of several to many meters. Improving understanding of ice dynamics is a high priority for scientific research to improve sea level rise projections.

Patterns consistent among different climate change models have led to some common expectations: GHG-induced climate change would include more heat waves and fewer extreme cold episodes; more precipitation on average but more droughts in some regions; and generally increased summer warming and dryness in the central portions of continents. Regional changes may vary from the global average changes, however. Scientists also expect precipitation to become more intense when it occurs, thereby increasing runoff and flooding risks.

Precipitation is a particularly challenging component of projecting future climate. For example, for the contiguous United States, recent climate modeling consistently anticipates overall temperature increases, but different models produce a wide range of precipitation changes, from net decreases to net increases.[19] This is particularly problematic in that precipitation, and its characteristics, is closely associated with impacts on agriculture, water supply, streamflows, and other critical systems.

Scientific expectations and model projections consistently point to a global average increase in precipitation with strong variations across regions and time. Generally, dry areas are expected to get dryer, and wet regions are expected to get wetter. In many regions, the increase in evapotranspiration is expected to exceed the increase in precipitation, resulting in general drying of soils and increasing risks of droughts. Precipitation, when it occurs, is expected to be more intense. There will be more energy available for storms, including hurricanes and thunderstorms, though whether they may increase in frequency remains unclear. Sea ice cover in the Arctic is projected to continue its recent decline (*Figure 5*). Greenland is expected to continue ice loss, adding to sea level rise, with more uncertainty about what may happen to ice cover in Antarctica (*Figure 5*). Because Arctic sea ice already floats on water, its melting would not increase sea levels, but large scale melting of land-based ice in Greenland and Antarctica could increase average sea levels by as many as 2 meters by 2100 and several more meters over coming centuries.

IMPACTS OF CLIMATE CHANGE

Nearly every human and natural system could be affected by climate changes, directly or indirectly. The U.S. Global Change Research Program has produced several assessments of scientific understanding of impacts of climate change on the United States.[20]

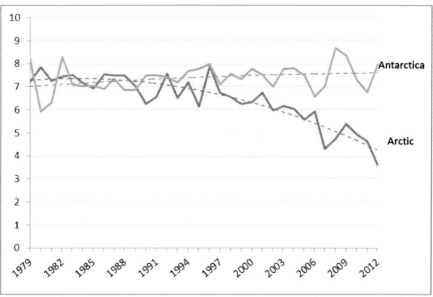

Source: CRS figure from data at the National Snow and Ice Data Center (extracted March 29, 2013), at http://nsidc.org/data/docs/noaa/g02135_seaice_index/ #monthly_graphs_format.

Notes: The dotted lines show the best polynomial fit (2-order) to each time series, as estimated by Excel. For both series, the polynomical fit was slightly better than a linear fit. See the National Snow and Ice Data Center website, referenced above, for further description of the underlying data.

Figure 5. Sea Ice Extent in Arctic (September) and Antarctica (April), 1979 to 2012. (in millions of square kilometers)

Climate Changes Would Affect A Wide Set of Human Systems

Changes in patterns of temperature, precipitation, sea levels, storms, and heat waves (among other indicators of climate) would affect, among other systems:

- water resources and delivery;
- agricultural productivity;
- the frequency and intensity of extreme weather events;
- spread of infectious diseases; air and water pollution levels;
- reliability of transportation, energy, and coastal protection systems;

- commodity prices;
- insurance pay-outs; and
- migration of people and species.

There are many additional elements of the economy and society that could be affected by shifts in climate. Research on potential impacts of climate change is generally less funded and developed than on the climate system itself.

Whether climate changes are meaningful in a policy context arguably depends, on the one hand, on how they influence existing and emerging human systems, and on the other hand, the values people attach to different resources and risks. Past climate changes, often regional not global, contributed to major societal changes, including some large-scale migrations and even the demise of some civilizations.[21] Some climate changes likely stimulated technological advances, such as development of irrigation systems.

Many investments in current buildings, transportation, water systems, agricultural hybrid varieties, and other infrastructure were designed in the context of a climate of one or more decades ago, cooler on average than today. To the degree that climate patterns were factored into design, the investments typically presumed that climate would remain stable within historical bounds of variability. For example, levees may have been built to withstand a 100-year flood (1% chance to occur each year) according to historic runoff, streamflow, and storm surge conditions going back many decades. As climate changes produce greater, more intense precipitation and run-off, however, a 100-year flood may now be closer to the 50-year flood (2% chance flood), and potentially the 10-year flood within decades (10% change flood). If climate continues to change from the conditions for which infrastructure and practices were designed, the risks of losses due to maladaptation would increase.

A wide band of uncertainty surrounds projections of impacts of climate change and, in particular, the critical thresholds for non-linear or abrupt effects. Some impacts of climate change are expected to be beneficial in some locations with a few degrees of warming (e.g., increased agricultural productivity in some regions, less need for space heating, less cold weather mortality, opening of the Northwest Passage for shipping and resource exploitation). Most impacts are expected to be adverse (e.g., lower agricultural productivity in many regions, drought, rising sea levels, spread of disease vectors, greater needs for cooling). Many impacts may be substantial but hard to assess as yet as "positive" or "negative," such as possible impacts on the structure of global financial markets. Risks of abrupt, surprising climate changes with accompanying dislocations are expected to increase as global average temperature increases; some could push natural and socioeconomic systems past key thresholds of tolerance.

Risks of future climate change would be reduced by efforts that reduce vulnerability and build resilience ("adaptation"). Some populations will have the resources to migrate and adapt successfully—even profit from new opportunities that will emerge—while others could lose livelihoods or lives. Adaptations can help mitigate impacts and damage costs, but also impose costs, often on those who can least afford them. Climate change will occur with different magnitudes and characteristics in different regions. The difficulties involved in improving predictions at regional and local scales will challenge preparations for climate change. To a large degree, climate change will expand the uncertainties that individuals and organizations face.

Climate change could have a wide array of effects on individuals, communities, and populations on a large scale. Many of these are expected to occur in small increments: shortages and increasing prices for clean water, rising food prices, higher rates of allergies and such illnesses as diarrhea or cholera, erosion of beaches, etc.

At an increasing rate may be shocks, or distinct weather events, such as more extreme heat waves, severe droughts, or loss of industrial cooling systems when intake water is in short supply or is warmer than tolerable temperatures.[22]

Atmospheric Carbon Dioxide is Increasing Ocean Acidity

The *acidity of the surface waters of the oceans* has increased by about 26% over the past 150 years.[23] Ocean acidification has occurred along with the rise in atmospheric concentrations of CO_2. The oceans remove 25%-40% of the carbon dioxide emissions added annually to the atmosphere by burning fossil fuels. The carbon dioxide absorbed in the oceans decreases the water's pH, an indicator of increasing acidity. According to a National Research Council (NRC) report, the current rate of acidification "exceeds any known change in ocean chemistry for at least 800,000 years."[24] Research shows varying sensitivities of different marine species to rising acidity, making general statements about impacts of ocean acidification difficult.

The NRC concluded,

> While the ultimate consequences are still unknown, there is a risk of ecosystem changes that threaten coral reefs, fisheries, protected species, and other natural resources of value to society. (*Executive Summary*, pp. 3-4)
>
> Congress enacted the Federal Ocean Acidification Research and Monitoring Act of 2009 (P.L. 111-11, Section 12311, Subtitle D) to improve monitoring and research, to assess carbon storage in the oceans and potential effects on acidification and other ocean conditions, and to develop predictive models for future changes in ocean chemistry and marine ecosystems. The program is housed within the National Oceanic and Atmospheric Administration (NOAA), and coordinated with other agencies through an interagency plan through the National Ocean Council.

Extreme events, chronic economic losses, or improved opportunities elsewhere are expected to prompt migration of millions of people, largely within countries, but also across national borders. Extreme events, greater variability, and uncertainty are expected to increase stress and mental health challenges. Some experts project that climate changes could amplify instabilities in countries with weak governance and increase security risks.[25] This may have implications for international political stability and security.[26]

For some experts and stakeholders, likely ecological disruptions (and limitations on species' and habitats' abilities to adapt at the projected rate of climate change) are among the most compelling reasons that humans must act to reduce their interference with the climate system. Some believe humans will have the wherewithal to cope, but non-human systems may not. As the degree and distribution of climate changes continue, ranges of species are likely to change. Climate change is highly likely to create substantial changes in ecological systems and services[27] in some locations, and may lead to ecological surprises.[28] The disappearance of some types of regional ecosystems raises risks of extinctions of species, especially those with narrow geographic or climatic distributions, and where existing ecological communities disintegrate.[29] One set of researchers found "a close correspondence between regions with globally disappearing climates and previously identified biodiversity hotspots; for these regions, standard conservation solutions (e.g., assisted migration and networked reserves) may be insufficient to preserve biodiversity."[30]

SELECTED, RELATED CRS REPORTS

CRS Report R43185, *Ocean Acidification*, by Harold F. Upton and Peter Folger.

CRS Report R41153, *Changes in the Arctic: Background and Issues for Congress*, coordinated by Ronald O'Rourke.

CRS Report RL34580, *Drought in the United States: Causes and Issues for Congress*, by Peter Folger, Betsy A. Cody, and Nicole T. Carter.

CRS Report R43199, *Energy-Water Nexus: The Energy Sector's Water Use*, by Nicole T. Carter.

CRS Report R42611, *Oil Sands and the Keystone XL Pipeline: Background and Selected Environmental Issues*, coordinated by Jonathan L. Ramseur.

CRS Report R42756, *Energy Policy: 113th Congress Issues*, by Carl E. Behrens.

CRS Report R42613, *Climate Change and Existing Law: A Survey of Legal Issues Past, Present, and Future*, by Robert Meltz.

CRS Report R43120, *President Obama's Climate Action Plan*, coordinated by Jane A. Leggett. CRS Report R42756, *Energy Policy: 113th Congress Issues*, by Carl E. Behrens. CRS Report RL34266, *Climate Change: Science Highlights*, by Jane A. Leggett.

CRS Report R41973, *Climate Change: Conceptual Approaches and Policy Tools*, by Jane A. Leggett.

End Notes

[1] This CRS report will be reviewed and, as appropriate, revised considering evidence provided from emerging scientific research. Of, note, the Intergovernmental Panel on Climate Change (IPCC) will release its fifth assessment report later in 2013.

[2] For more information, see maps available at the National Climate Data Center, http://www.ncdc.noaa.gov/oa/climate/globalwarming.html and http://www. ncdc.noaa.gov /oa/climate/research/trends.html#global.

[3] Board on Atmospheric Sciences and Climate. Surface Temperature Reconstructions for the Last 2,000 Years. National Research Council, 2006. http://books.nap.edu/openbook.php? record_id=11676&page=1.

[4] Marcott, Shaun A., Jeremy D. Shakun, Peter U. Clark, and Alan C. Mix. "A Reconstruction of Regional and Global Temperature for the Past 11,300 Years." Science 339, no. 6124 (March 8, 2013): 1198–1201. doi:10.1126/science.1228026; Kellerhals, T., S. Brütsch, M. Sigl, S. Knüsel, H. W. Gäggeler, and M. Schwikowski. "Ammonium Concentration in Ice Cores: A New Proxy for Regional Temperature Reconstruction?" Journal of Geophysical Research: Atmospheres 115, no. D16 (2010): n/a–n/a. doi:10.1029/2009JD012603; Thibodeau, Benoît, Anne de Vernal, Claude Hillaire-Marcel, and Alfonso Mucci.

"Twentieth Century Warming in Deep Waters of the Gulf of St. Lawrence: A Unique Feature of the Last Millennium." Geophysical Research Letters 37, no. 17 (2010): n/a–n/a. doi:10.1029/2010GL044771. See also the references at http://www.globalwarmingart. com/wiki/File:Holocene_Temperature_Variations_Rev_png, which depict a collection of major temperature reconstructions of the Holocene, as well as the broad range of uncertainty of available estimates and the average of those estimates.

[5] National Research Council. Ocean Acidification: A National Strategy to Meet the Challenges of a Changing Ocean. Washington DC, 2013; Feely, Richard A. 2010. A Rational Discussion of Climate Change: The Science, the Evidence, the Response. Testimony before the House Committee on Science and Technology, Subcommittee on Energy and Commerce. Washington DC. (p.130). See also CRS Report R43185, Ocean Acidification, by Harold F. Upton and Peter Folger.

[6] See, for example, the discussion in Kanazawa, Satoshi. "Common misconceptions about science I: "Scientific proof"." Psychology Today, November 16, 2008. http://www. psychologytoday.com/blog/the-scientific-fundamentalist/ 200811/common-misconceptions-about-science-i-scientific-proof.

[7] Many CRS reports related to climate change may be found at Issues Before Congress: Climate Change Science, Technology, and Policy, at http://www.crs.gov/pages/subissue.aspx? cliid=3878&parentid=2522&preview=False.

[8] Water vapor is the most important GHG in the atmosphere but is understood not to be directly influenced by humans; it would be, however, involved in feedback mechanisms, discussed later.

[9] About one-third of human-related CO2 has been absorbed by oceans, increasing surface water acidity by 30%. See National Research Council. Ocean Acidification: A National Strategy to Meet the Challenges of a Changing Ocean. Washington, DC, 2013; Feely, Richard A. 2010. A Rational Discussion of Climate Change: The Science, the Evidence, the Response. Testimony before the House Committee on Science and Technology, Subcommittee on Energy and Commerce. Washington, DC. (p.130).

[10] One line of evidence is that carbon dioxide levels have varied closely with the Earth's temperature in and out of glacial periods over the past million years. These cycles are mostly triggered by changes in the Earth's orbit, tilt, and wobble. In some of these cycles, temperatures rose in advance of rising atmospheric carbon dioxide concentrations. Scientists generally interpret this as a tendency for positive climate warming feedbacks that increase carbon dioxide concentrations which then enhance warming, etc.—that the net positive feedbacks amplify an initial climate warming.

[11] Positive feedbacks could increase if and when, for example, large tracts of forests die back as a response to exceeding their climate thresholds of tolerance, or current permafrost thaws and releases the carbon it contains, or if reservoirs of methane hydrates destabilize.

[12] In order to show multiple gases of different potencies on a single scale, GHG have been indexed relative to the effect that a mass of CO2 would have over several time periods (because GHG remain in the atmosphere for different lengths of time, from days to tens of thousands of years). The index used for these estimates uses a 100-year time horizon, the most frequently used period.

[13] Intergovernmental Panel on Climate Change Working Group I, Climate Change 2007: The Physical Basis (Cambridge, UK: Cambridge University Press, 2007).

[14] As a point of reference, the global mean annual temperature during the 20th century is estimated to have been approximately 13.9o Celsius (57.0o Fahrenheit), according to NOAA's National Climate Data Center.

[15] Ibid.

[16] Intergovernmental Panel on Climate Change Working Group I. Climate Change 2007: The Physical Basis. Cambridge, UK: Cambridge University Press, 2007. Executive Summary.

[17] Highest temperatures of the Holocene may have occurred in one or more periods some 5,000 to 8,000 years ago, although sufficient data are not available for all parts of the globe to have reliable estimates of average global temperature. The oldest cities discovered date from approximately the same period, such as the extensive settlement of Byblos in present-day Lebanon, by about 6,000 years ago, or Medinat Al-Fayoum in Egypt, about 6,000 years old. Since the early to mid-Holocene, however, average temperatures appear to have been declining slowly, with notable periods of warming and cooling. The changes entailed in Holocene climate variability have been significant in terms of effects on humans and ecosystems, and have led to both benefits to, and the demise of, numerous civilizations.

[18] See discussion in National Research Council. Advancing the Science of Climate Change. Washington DC, 2010, at p. 244.

[19] See, for example, climate change scenarios available from the U.S. Global Change Research Program at http://scenarios.globalchange.gov/sites/default/ files/b/figures/ UnitedStates/ Ann_US_precip_a2.png, with notes at http://scenarios.globalchange.gov/node/1087. See also discussion in this report regarding dealing with uncertainties.

[20] Karl, Thomas R., Mellillo, Jerry M., and Peterson, Thomas C. (eds.) Global Change Impacts in the United States. U.S. Global Change Research Program. 2009. Such periodic assessments are required by the Global Change Research Act of 1990 (P.L. 101-606). A new national assessment of impacts on the United States is due in late 2013.

[21] There is a growing set of research on the relationship between past climate change and civilizations. A sample of recent research includes Buckley, B. M., K. J. Anchukaitis, D. Penny, R. Fletcher, E. R. Cook, M. Sano, L. C. Nam, A. Wichienkeeo, T. T. Minh, and T. M. Hong. "Climate as a Contributing Factor in the Demise of Angkor, Cambodia." Proceedings of the National Academy of Sciences 107, no. 15 (March 2010): 6748–6752; Cook, Edward R, Kevin J Anchukaitis, Brendan M Buckley, Rosanne D D'Arrigo, Gordon C Jacoby, and William E Wright. "Asian Monsoon Failure and Megadrought During the Last Millennium." Science (New York, N.Y.) 328, no. 5977 (April 23, 2010): 486– 489; DeMenocal, PB. "Cultural Responses to Climate Change During the Late Holocene." Science (Washington) 292, no. 5517 (April 27, 2001): 667–673; Haug, G. H, D. Gunther, L. C Peterson, D. M Sigman, K. A Hughen, and B. Aeschlimann. "Climate and the Collapse of Maya Civilization." Science 299, no. 5613 (2003): 1731; Scholz, Christopher A., Thomas C. Johnson, Andrew S. Cohen, John W. King, John A. Peck, Jonathan T. Overpeck, Michael R. Talbot, et al. "East African Megadroughts Between 135 and 75 Thousand Years Ago and Bearing on Early-modern Human Origins." Proceedings of the National Academy of Sciences of the United States of America 104, no. 42 (October 16, 2007): 16416–16421.

[22] For examples of these risks to power plants, see Department of Energy (DOE), U.S. Energy Sector Vulnerabilities to Climate Change and Extreme Weather. July 2013. http://energy.gov/sites/prod/files/2013/07/f2/20130710-EnergySector-Vulnerabilities-Report.pdf. See also CRS Report R43199, Energy-Water Nexus: The Energy Sector's Water Use, by Nicole T. Carter.

[23] NRC Committee on the Development of an Integrated Science Strategy for Ocean Acidification Monitoring, Research, and Impacts Assessment; National Research Council. "Executive Summary." In Ocean Acidification: A National Strategy to Meet the Challenges of a Changing Ocean. Prepublication. Washington, D.C.: The National Academies Press, 2010; Jacobson, Mark Z. "Studying ocean acidification with conservative, stable numerical

schemes for nonequilibrium air-ocean exchange and ocean equilibrium chemistry." Journal of Geophysical Research 110 (April 2, 2005): 17 PP.

[24] Ibid.

[25] An example of this is the adverse weather events in early 2011 that led to spikes in food prices and contributed to demonstrations in Tunisia and Egypt. These, in turn, led to regime change, although one cannot attribute these events to climate change, as opposed to weather variability, and the political implications might have been very different in regimes with better economic performance, less income disparity, fewer allegations of corruption, and greater social resilience. The point remains, nonetheless, that societies are sensitive to climatic variables in many ways.

[26] Regarding risks to national security, see, for example, Defense Science Board Task Force on Trends and Implications of Climate Change for National and International Security. October 2011. http://www.acq.osd.mil/dsb/ reports/ADA552760.pdf; and U.S. Department of Defense. Quadrennial Defense Review, February 2010. (pp. xv, 84- 88) http://www. defense.gov/qdr/qdr%20as%20of%2029jan10%201600.pdf.

[27] Economists and scientists sometimes refer to "ecosystem services," which are the services that natural systems provide and for which, very frequently, humans do not typically pay. Ecosystems services include water filtration, filtering of air pollution, recreational and spiritual opportunities, etc. Even without being valued in capital markets, ecosystem services may be critically important to economies. For example, in many coastal areas, mangroves or wetlands provide valuable buffering against frequent storm and flood events. If such ecosystem services did not exist, communities would have to pay for manufactured alternatives (e.g., sea walls) or risk incurring damages.

[28] For example, the very rapid spread of pine beetles in recent years was unexpected and caused large damages (although a temporarily inexpensive supply of timber) in a very short period. See CRS Report R40203, Mountain Pine Beetles and Forest Destruction: Effects, Responses, and Relationship to Climate Change.

[29] Malcolm, Jay R., Canran Liu, Ronald P. Neilson, Lara Hansen, and Lee Hannah. "Global Warming and Extinctions of Endemic Species from Biodiversity Hotspots." Conservation Biology 20, no. 2 (2006): 538-548.

[30] John W. Williams, Stephen T Jackson, and John E. Kutzbach, "Projected distributions of novel and disappearing climates by 2100 AD," Proceedings of the National Academy of Sciences of the United States of America 104, no. 14 (April 3, 2007).

INDEX

#

20th century, 77, 82, 92
21st century, 85

A

access, 64, 71
accessibility, 34
acidity, 78, 89, 92
AD, 94
adaptation, vii, 1, 2, 3, 4, 8, 9, 10, 11, 27,
    28, 31, 40, 53, 54, 55, 56, 57, 58, 59, 62,
    63, 64, 69, 72, 73, 89
adverse effects, viii, 76
adverse weather, 94
aerosols, 78, 81, 82
aesthetics, 8
age, 7, 67, 68
agencies, 36, 64, 68, 70, 89
agriculture, 78, 81, 86
air quality, 30, 71
Alaska, 4
alternative causes, 79
appropriations, 80
assessment, 91, 93
assets, 55, 61
atmosphere, 78, 80, 81, 83, 84, 89, 92
atmospheric deposition, 69, 71
attribution, 79

authorities, 80
avian, 68

B

basic research, 38
beetles, 21, 47, 94
benefits, vii, viii, 4, 22, 76, 78, 93
benign, 37
bias, 56
biodiversity, 8, 90
biotic, 7
birds, 6, 7
bleeding, 32
boils, 59
bounds, 83, 88

C

Cambodia, 93
capital markets, 94
carbon, 52, 78, 81, 84, 85, 89, 92
carbon dioxide, 52, 78, 81, 85, 89, 92
challenges, 2, 11, 29, 31, 35, 38, 39, 40, 50,
    58, 60, 90
China, 81
cholera, 89
cities, 43, 93
City, 67
classes, 67, 68

96 Index

cleanup, 21
climate change impacts, vii, 1, 2, 3, 4, 6, 8, 10, 27, 28, 31, 40, 51, 53, 54, 55, 58, 59, 62, 63, 73, 78
climates, 90, 94
CO_2, 71, 78, 81, 83, 84, 89, 92
collaboration, 53, 64
combined effect, 7
commodity, 87
communities, 7, 18, 28, 29, 31, 72, 89, 90, 94
community, 19, 29
competition, 37
complexity, 8
compliance, 39
composition, 2, 42, 43, 45, 50, 51, 80
computer, 24, 26
conflict, 14, 29, 30
congress, 18, 20, 72, 75, 80, 89, 91, 92
conifer, 68, 70
connectivity, 8
consensus, 78
conservation, 56, 90
consumers, 49
continental, 70
controversial, vii, 10, 75
controversies, 76
convention, 4
cooling, 77, 88, 89, 93
cooperation, 29
coral reefs, 89
corruption, 94
Costa Rica, 6
crop, 37
crust, 68
cycles, viii, 2, 48, 49, 50, 75, 92
cycling, 69

D

damages, 80, 94
danger, 59
data collection, 2, 3, 4, 17, 22, 35, 40, 55, 59, 60, 61, 62, 63
database, 54, 59

decomposition, 52
deforestation, 78, 81
degradation, 81
demonstrations, 94
Department of Agriculture, 1
Department of Defense, 94
Department of Energy, 93
deposition, 70
depth, 2, 10, 26, 63
diarrhea, 89
diseases, 21, 44, 50, 87
distribution, 5, 8, 46, 56, 90
diversification, 36
diversity, 8, 55, 63
downsizing, 59
draft, 66
drawing, 47
drought, 7, 8, 9, 45, 47, 51, 88
drying, 45, 86
dynamic systems, 42, 49

E

echoing, 58
ecological information, 9
ecological systems, 26, 85, 90
ecology, 6, 17, 20, 23, 25, 38, 43, 44, 51, 67, 68, 69
economic losses, 90
economic performance, 94
economics, viii, 76, 79
ecosystem, 6, 7, 8, 9, 10, 13, 22, 26, 50, 55, 69, 71, 89, 94
ecosystems, vii, viii, 2, 6, 8, 9, 22, 24, 45, 49, 50, 51, 54, 55, 62, 63, 69, 72, 76, 78, 79, 84, 89, 90, 93
Egypt, 93, 94
El Niño, 84
emission, 84
employees, vii, 1, 2, 3, 4, 9, 10, 11, 16, 21, 22, 23, 27, 28, 29, 31, 34, 35, 39, 41, 42, 44, 45, 48, 49, 50, 51, 59, 60, 62, 63, 64
employment, 22
energy, 71, 78, 81, 86, 87, 93

environment(s), 9, 12, 13, 27, 42, 55, 56, 58, 60, 63, 64, 68, 69
environmental aspects, 31
environmental impact, 39
Environmental Protection Act, 40
environmental stress, 7
equilibrium, 94
erosion, vii, 5, 30, 89
evapotranspiration, 86
evidence, viii, 6, 9, 11, 76, 79, 91, 92
Experimental Forests and Ranges, i, iii, v, vii, 1, 2, 3, 4, 11, 33, 65, 66, 67, 73
exploitation, 88
extreme cold, 86
extreme weather events, 87

F

fertilization, 37, 78
fidelity, 52
filtration, 94
financial, 59, 80, 89
financial incentives, 80
financial markets, 89
fire dynamics, vii, 5, 6
fire hazard, 29
fires, 36, 44, 45, 46, 50
fisheries, 89
fishing, 29
flooding, 69, 86
floods, 18
fluctuations, 60, 85
food, 44, 89, 94
force, 49, 81
forest management, 30, 35, 69
forest regeneration, vii, 5
forest restoration, 69
freedom, 33
funding, 3, 15, 32, 33, 34, 35, 36, 37, 38, 40, 41, 59, 60, 61, 63, 64, 72
funds, 3, 32, 63

G

genetics, 17, 68, 70
GHG, vii, 75, 76, 78, 79, 81, 82, 83, 84, 85, 86, 92
Global Change Research Act, 93
global climate change, 6
global scale, 5
global warming, viii, 76, 78
governance, 90
grants, 32
grass, 32
grasslands, 70
grazing, 18, 29, 68, 69
greenhouse, vii, 75, 79, 81
greenhouse gas (GHG), vii, 75, 79, 81
greenhouse gases, vii, 75, 79
growth, 15, 38, 42, 43, 46, 47, 50, 57, 68, 69, 70, 83

H

habitat(s), 2, 47, 50, 55, 67, 69, 90
Hawaii, 4
hemisphere, 68
herbicide, 29
hiring, 33
history, 2, 10, 16, 22, 37, 38, 45, 56, 58, 63, 67, 68, 77
Holocene, 92, 93
homes, 29
homogeneity, 60
host, vii, 1, 6, 12, 17, 19, 53, 58, 70
hotspots, 90
House, 92
human, viii, 13, 18, 42, 60, 76, 78, 79, 81, 85, 86, 88, 90, 92
human nature, 60
hunting, 29, 30
hurricanes, 86
hybrid, 88
hydrology, vii, 5, 6, 16, 51, 52
hypothesis, 20, 56, 79

Index

I

ideal, 48, 54
improvements, viii, 76
incidence, 46
income, 81, 94
indirect effect, 47
individuals, 89
industry, 78, 81
infestations, 47, 48
infrastructure, 15, 32, 40, 41, 59, 88
initiation, 47
insects, 21, 43, 44, 45, 48, 50
institutional change, 3, 28, 31, 40, 72
institutions, 64
integrity, vii, 1, 2, 12, 14, 15, 27, 54, 55, 61, 62, 64
interface, 38
interference, 30, 62, 90
investment(s), 3, 35, 61, 64, 88
irrigation, 88
islands, 55
isolation, 51
issues, 17, 35

J

Japan, 6

K

kill, 36, 47, 51
Kyoto Protocol, 83, 84

L

land cover change, viii, 75, 78
land management, vii, 4, 9, 15, 18, 19, 70
land managers, vii, 4, 9, 19, 27, 72
land owners, vii, 4
landscape, 8, 9, 12, 14, 20, 21, 54, 58, 68
landscapes, 12, 13, 21, 27, 38, 54, 58, 62, 63, 67

law enforcement, 39, 40
lead, 57, 83, 85, 90
leadership, 2, 19, 35, 61, 63
leaks, 81
learning, 24
Lebanon, 93
levees, 88
light, 80
local community, 28
logistics, 23
love, 24, 57

M

majority, viii, 10, 22, 55, 76, 79, 81
mammal, 68
management, vii, 2, 3, 4, 8, 9, 12, 14, 15, 17, 18, 19, 20, 22, 23, 27, 29, 34, 37, 38, 39, 40, 51, 53, 56, 57, 62, 63, 67, 68, 69, 70
mangroves, 94
manipulation, 56
marine species, 89
mass, 92
mathematics, 79
matter, 59
measurement(s), 15, 25, 26, 32, 62, 79
media, 76
melting, 85, 86
memory, 33
mental health, 90
Mexico, 13, 52
migration, 8, 9, 87, 90
misconceptions, 92
models, 86, 89
momentum, 71
Montana, 5, 10, 66, 67, 68, 70
mortality, 7, 43, 46, 69, 88

N

National Academy of Sciences, 77, 93, 94
national borders, 90

Index

National Forest System, vii, 2, 8, 13, 28, 39, 57, 68
National Research Council, 89, 91, 92, 93
national security, 94
National Strategy, 92, 93
native species, 2, 42, 45, 50
natural climate variability, viii, 75, 78
natural resource management, 63
natural resources, 13, 89
nitrous oxide, 78, 81
NOAA, 89, 92
nonequilibrium, 94
North America, 64, 71
NRC, 89, 93
nutrient, 69

O

oceans, viii, 75, 78, 89, 92
opportunities, 2, 3, 11, 12, 13, 20, 21, 27, 31, 38, 39, 40, 41, 50, 52, 60, 67, 70, 89, 90, 94
orbit, 78, 81, 92
organize, 11
oscillation, 84
outreach, 28
outreach programs, 28
ozone, 81

P

Pacific, 84
parasites, 44
participants, vii, 1, 2, 3, 10, 11, 12, 14, 15, 16, 17, 18, 19, 21, 28, 30, 31, 33, 34, 35, 36, 37, 40, 41, 42, 44, 48, 49, 50, 53, 54, 55, 56, 57, 58, 59, 60, 61, 65, 69, 74
pathogens, 2, 7, 48, 50
pathology, 69
permafrost, 92
pH, 89
plants, 6, 43, 45
playing, 59
policy, 88

pollutants, vii, 75
pollution, 87, 94
pools, 64
population, 8, 48, 49, 68
positive feedback, 83, 92
power plants, 93
precedent, 57
precipitation, viii, 6, 7, 45, 48, 49, 50, 51, 52, 68, 76, 83, 85, 86, 87, 88
predators, 44, 49
President, 91
President Obama, 91
probe, 72
professionals, 33
profit, 89
project, vii, viii, 1, 9, 10, 11, 14, 23, 51, 52, 65, 70, 76, 79, 85, 90
protected areas, 8
protection, 2, 14, 18, 27, 39, 40, 87
public education, 28
public interest, 36
Puerto Rico, 4

R

radiation, viii, 76, 81
range management, vii, 4, 17, 27, 69
rangeland, 68
reading, 26
reality, 4
recall, 10
recognition, 9, 31, 34, 35, 41, 60
recommendations, 63
recovery, 38
recreation, 28, 29, 30, 31, 68
recreational, 28, 29, 30, 31, 40, 94
redistribution, 52
regenerate, 18
regeneration, vii, 5, 18, 42, 68, 69
regulations, 28
relevance, 2, 11, 17, 18, 63
reliability, 87
relief, 29
representativeness, 27, 53, 54, 55, 61
requirements, 56

Index

researchers, 10, 17, 26, 27, 33, 51, 62, 63, 68, 90
reserves, 90
resilience, 8, 9, 38, 53, 55, 62, 89, 94
resistance, 8
resources, 8, 29, 31, 36, 41, 64, 88, 89
response, 43, 53, 56, 69, 80, 92
restoration, 36, 38, 68, 69
restoration plantings, 69
restrictions, 30, 40
risk(s), viii, 76, 79, 80, 85, 86, 88, 89, 90, 93, 94
RNA(s), 57
Rocky Mountain Research Station, vii, 1, 2, 5, 10, 28, 32, 39, 65, 66, 67, 73
rodents, 49
root(s), 25, 32, 37
root system, 37
routes, 23
rules, 28
runoff, 86, 88

S

scarcity, 60
school, 69
science, vii, 2, 12, 16, 17, 19, 22, 27, 33, 41, 61, 62, 63, 64, 75, 76, 78, 79, 80, 85, 91, 92
scientific knowledge, 9, 23
scientific method, viii, 76, 79
scientific theory, 79
scientific understanding, 82, 86
scope, viii, 76, 78
sea level, vii, 75, 76, 85, 86, 87, 88
sea level rise, vii, 75, 76, 85, 86
security, 90
seed, 51, 68, 69
seeding, 68
seedlings, 48, 56
semi-structured interviews, 10
sensitivity, 31
services, 94
shade, 42
sheep, 68

shoots, 46
short supply, 13, 89
shortfall, 3, 63
showing, 6, 22
shrubs, 70
Siberia, 6
signs, 25
silviculture, vii, 5, 18, 20, 69, 70
smog, 30
social context, 31, 53
society, 37, 87, 89
software, 11
soil type, 69
South America, 71
South Dakota, 5, 67
species, 2, 6, 7, 8, 42, 43, 44, 45, 47, 50, 51, 55, 56, 62, 72, 87, 89, 90
stability, vii, 5, 58, 59, 90
staff members, 10, 22
staffing, 72
stakeholders, 90
state(s), 7, 8, 21, 64, 70, 85
storage, 34, 89
storms, viii, 45, 76, 86, 87
stress, 47, 90
structure, 3, 42, 47, 50, 89
succession, 41, 42, 45, 57
sulfur, 81
Sun, vii, 75, 81

T

Task Force, 94
technician, 23, 24, 33
technological advances, 88
technology, 80
teeth, 17
temperature, 7, 46, 51, 68, 77, 78, 79, 82, 85, 86, 87, 89, 92, 93
tension(s), 13
tenure, 10
territory, 13, 68
testing, 20, 56, 73
thinning, 9, 39, 40, 41, 56, 69
timber production, 37, 38, 41, 69

Index

time periods, 59, 92
time series, 77, 87
trade, 60
trade-off, 60
traits, 58
transportation, 87, 88
treatment, 12, 29, 38, 52
tundra, 69

U

U.S. Department of Commerce, 77
UK, 92, 93
uniform, 77
United Kingdom, 6, 7
United Nations, 6
United States, 1, 4, 7, 16, 18, 71, 77, 81, 84, 86, 91, 93, 94
universities, 32, 64, 68, 69, 70
urban, 38, 70, 81
USDA, 3, 4, 65, 66, 67
USGS, 61

V

vapor, 92
variables, 60, 94
variations, 25, 48, 85, 86

varieties, 88
vegetation, viii, 6, 20, 37, 43, 44, 51, 52, 56, 57, 67, 68, 69, 75, 78, 81, 83
vulnerability, 89

W

walking, 23
Washington, 65, 67, 92, 93
water, 13, 15, 17, 69, 70, 71, 78, 81, 83, 86, 87, 88, 89, 92, 94
water chemistry, 69
water quality, 70
water resources, 87
water vapor, 78, 81
watershed, 23, 37, 69, 70
wear, 24
weather patterns, 50
wetlands, 94
wilderness, 21
wildfire, 44, 46
wildland, 13, 28, 29, 68, 69, 70
wildland-urban interface, 13, 28, 29, 68, 69, 70
wildlife, 2, 38, 42, 45, 48, 49, 50, 51, 70
windstorms, 7
worldwide, 16